THE THIRD

States of Mind and Being

TITLES OF RELATED INTEREST

African environments and resources
L. Berry & L. Lewis

Black Africa 1945–1980
D. Fieldhouse

The city in cultural context
J. Agnew *et al.* (eds)

Coping with hunger
P. Richards

Development economics and policy
I. Livingstone (ed.)

East Indians in a West Indian town
C. Clarke

Environment and development
P. Bartelmus

Environmental change and tropical geomorphology
I. Douglas & T. Spencer (eds)

A geography of Brazilian development
J. D. Henshall & R. P. Momsen jr

Geology and mineral resources of West Africa
J. B. Wright

Geomorphology in arid regions
D. O. Doehring (ed.)

The Green Revolution revisited
B. Glaeser (ed.)

Industrial structure and policy in the less developed countries
C. H. Kirkpatrick *et al.*

Interpretations of calamity
K. Hewitt (ed.)

International capitalism and industrial restructuring
R. Peet (ed.)

Introduction to development economics
S. Ghatak

Introduction to the politics of tropical Africa
R. Hodder-Williams

Learning from China?
B. Glaeser (ed.)

Living under apartheid
D. Smith (ed.)

Political change in the Third World
C. Andrain

Politics, security and development in small states
C. Clarke & A. Payne (eds)

The price of war
D. Forbes & N. Thrift

Regional development and settlement policy
R. Dewar *et al.*

Rich and poor countries
H. W. Singer & J. A. Ansari

Science, technology and social change
S. Yearley

Shelter, settlement and development
L. Rodwin (ed.)

The sociology of modernization and development
D. Harrison

Technology and rural women
L. Ahmed (ed.)

Third World proletariat?
P. Lloyd

Welfare politics in Mexico
P. Ward

THE THIRD WORLD

States of Mind and Being

Edited by

JIM NORWINE

Department of Geosciences, Texas A & I University

ALFONSO GONZALEZ

Department of Geography, University of Calgary

Boston

UNWIN HYMAN

London Sydney Wellington

Unwin Hyman, Inc.,
8 Winchester Place, Winchester, Mass. 01890, USA

Published by the Academic Division of
Unwin Hyman Ltd
15/17 Broadwick Street, London W1V 1FP, UK

Allen & Unwin (Australia) Ltd,
8 Napier Street, North Sydney, NSW 2060, Australia

Allen & Unwin (New Zealand) Ltd in association with the
Port Nicholson Press Ltd,
60 Cambridge Terrace, Wellington, New Zealand

First published in 1988

Library of Congress Cataloging-in-Publication Data

The Third World: states of the mind and being / edited by Jim Norwine,
Alfonso Gonzalez.
p. cm.
Includes bibliographies and index.
ISBN 0-04-910106-4 (alk. Paper)
0-04-910121-8 (pbk.: alk. paper)
1. Developing countries—Economic conditions. 2. Natural resources—Developing countries. 3. Developing countries—Social conditions. 4. Developing conditions—Politics and government.
I. Norwine, Jim. II. Gonzalez, Alfonso, 1927–

HC59.7.T465 1988
330.9172'4—dc19

British Library Cataloguing in Publication Data

The Third World: states of mind
and being
1. Developing countries
I. Norwine, Jim 1943– . II. Gonzalez Alfonso
909'.097240828

ISBN 0-04-910106-4
ISBN 0-04-910121-8 Pbk

Typeset in 10 on 12 point Bembo by
Computape (Pickering) Ltd, Yorkshire
and printed in Great Britain by
Biddles of Guildford

Gratefully, to Balfie, Big Ben, Nur Quddus, and Beau Stanley, to the memory of Captain Hook and – always – to you, Zaa: pearls may be forgotten, but never their casters.

> Out beyond ideas of rightdoing and wrongdoing there is a field.
> I'll meet you there. (J. Rumi)

Acknowledgements

Most of the credit – blame? – for this foolhardy attempt at defining the undefinable must go to our geography students at Texas A & I University and the University of Calgary. To their voracious and insatiable curiosity about the nature and significance of Earth's socioeconomic and sociocultural "worlds"; to their refusal to accept our comfortable paradigms; and to the touchstone they provided in the forms of their visceral insistence on focusing on the "truisms" of global patterns of human systems – patterns, of course, equally vital, ephemeral, and misleading – we owe the inspiration and exasperation which prompted this work. Faculty and administrative colleagues at our respective universities also offered invaluable counsel, for which we are most grateful. Most of all, we are grateful to our families, especially our wives, for support and forbearance above and beyond the call of duty.

We gladly acknowledge these debts. We solely are responsible for any shortcomings which might remain in the book.

Contents

Introduction

> I don't know if any of you have noticed, early in the morning, the sunlight on the waters. How extraordinarily soft is the light, and how the dark waters dance, with the morning star over the trees, the only star in the sky. Do you ever notice any of that? Or are you so busy, so occupied with the daily routine, that you forget or have never known the rich beauty of this earth – this earth on which all of us have to live? Whether we call ourselves communists or capitalists, Hindus or Buddhists, Moslems or Christians, whether we are blind, lame, or well and happy, this earth is ours. Do you understand? It is our earth, not somebody else's; it is not only the rich man's earth, it does not belong exclusively to the powerful rulers, to the nobles of the land, but it is our earth, yours and mine. We are nobodies, yet we also live on this earth, and we all have to live together. It is the world of the poor as well as of the rich, of the unlettered as well as of the learned; it is *our* world, and I think it is very important to feel this and to love the earth, not just occasionally on a peaceful morning, but all the time.
>
> J. Krishnamurti, *Think on These Things* (1970)

The Third World, like distant stars, should be studied obliquely: elusive at best, under direct observation it appears to dissolve so readily that one questions whether, indeed, it was ever really there.

The specific definition or meaning of the "Third World" is highly debatable, as we shall see. The greater problem is that we can't even agree whether or not such a place exists at all. Some hold that three worlds are two too many: that ultimately all human beings live in and experience the same single world. Others take quite a different stance, arguing that a global socioeconomic classification scheme with only three or four categories is far too simplistic and generalized properly to reflect the great diversity of cultures, economies and values across the globe. The social commentator Shiva Naipaul, for example, in one of the last essays of a brilliant but tragically short career, claimed that so great is the internal diversity of the "Third World" that its existence can be no more than mythical: "there is no such thing as the 'Third World'" (Naipaul 1985).

All that is true enough: in a literal sense there *is* only one world and Shiva Naipaul is certainly also right, the Third World is nothing if not heterogeneous (including as it does the world's wealthiest and poorest nations per capita). And one must finally also concede that the very idea of the Third World was devised – and remains most popular – in the First World. And yet, so what?

The Third World is "mythical" in that it is not "real" – in the sense of being concrete or tangible – but is rather an abstraction, a mental construct, an idea. It is in that way similar to numbers, which are, as Kenneth Boulding has put it, "extremely useful figments of the human imagination" (Boulding 1980).

The concept "Third World" actually is at once intellectual, metaphysical and (this can hardly be stressed too much) *experiential*; thus, it is much more than merely a helpful but fundamentally trivial – or even nonexistent – idea. It *is* a mental region – an image, if you wish – but a tremendously profound and vital image.

The mental-map region we may term the "Third World" is rather like a climatic region; say, a desert. One cannot "see" a desert but only such inferential clues as cacti, wadis and rattlesnakes. Where, for instance, does the Sahara begin or end? One cannot give a precise answer, yet most of us intuit that the Sahara does exist.

Consider, for a sociocultural example, "Macondo", the mythical Colombian town of Gabriel García Márquez' epic novel, *One Hundred Years of Solitude*. "Macondo, as a town, doesn't exist *because it exists everywhere* [our emphasis]. It is a part of the consciousness of South America" (Dreiske 1986, García Márquez 1971).

Shiva Naipaul argues that the internal diversity of the Third World is so extreme as to vitiate the very idea of such a region (mental or otherwise). Naipaul misses a fundamental point: some regions are only definable or distinguishable, perversely enough, through this extreme internal diversity. A fine example in the biophysical world is the tropical rainforest. More diverse in flora and fauna than any other terrestrial biogeographic type, a rainforest is nonetheless one organic whole (or even "individual," as Stephen Jay Gould puts it), consisting of many disparate parts, yet far greater than the sum of them. (Not unlike individual human beings, actually. . . .)

One intuits the Third World via a distinctive – vis-à-vis the First World – combination of perspectives (e.g., self-view and world-view), attitudes and values, and, in varying degrees, the relative absence of material expressions of affluence. Hence the subtitle of this book: states of mind *and* of being.

Many observers have addressed the notion of distinctive categories of human life experiences which differ from one another both quantitatively and qualitatively. That the categories' names vary from expert to expert is of little consequence other than, perhaps, the psychological: "rich" and "poor" (these being perhaps at once the most comprehensible and most flawed); "developed" and "developing"; "North" and "South"; and, of course, the First, Second, Third (etc.) Worlds.

How many such "worlds" are there, really? Who knows? It is probably a matter of opinion. As Octavio Paz intimated in the title of a recent book: "One Earth, Four or Five Worlds" (Paz 1984). The "Third World" exists, whatever we choose to call it. The more difficult question is, how can we understand it?

The Third World endeavors to address this question. It consists of a series of short, state-of-the-art essays written by leaders in their respective fields. The essays represent only a selected sample: no single volume could do justice to such a rich subject. They are subjective – e.g., both leftist and rightist political stances are taken – but not, it is hoped, arbitrary. Taken together, they present an impressionistic but revealing portrait of the Third World.

Development

The problem of socioeconomic development and the concomitant improvement of levels of living or the quality of life is certainly one of the outstanding issues confronting mankind. Virtually all aspects regarding the problem of development are subject to controversy and there appears to be little general agreement on anything. Nevertheless, we will endeavor to provide perspectives from a variety of specialists on different aspects of this ever-present critical problem.

Development does not appear to be a question of national endowment, as are natural resources or population size, but of capabilities, such as the utilization of resources, technology, and socioeconomic institutions. A definition of development would include the processes of more effective use of resources and increased efficiency in production and distribution, which result in a greater volume and diversity of goods and services for less human physical labor. Some argue that the distributional aspects within a society are of major, perhaps primary, concern. To these advocates, the redistribution of wealth or the more egalitarian distribution of income constitutes "true" development. To this same group, and to others, "true" or "real" development consists of "human rights." Human rights generally consist of (a) political and civil rights (i.e., civil liberties and political freedom; this is the concept that is probably most frequently used in the western democracies), and (b) socioeconomic rights (i.e., the right to health care, an adequate diet, education, decent housing, and even the right to employment and rights for women and minorities). In the western democracy these are considered generally to be highly desirable objectives of social justice and government policy rather than "rights." However, this aspect of human rights is most frequently advocated in general by the underdeveloped world and especially by radicals and marxist regimes where political and civil rights are about the most restricted anywhere.

The causes of the status of development in the underdeveloped regions is also a subject of great controversy. Some groups, and even specialists, expound on one particular cause or another, including the social system and structure, culture, limited education and technology, inadequate resources, population growth and pressure, inadequate capital accumulation, colonialism/neocolonialism/exploitation, dependency on other countries or source regions, sociocultural isolation, the capitalist system (nationally and/or internationally), restraints on private enterprise/open market, and so forth. Despite the conventional wisdom that development is a complex issue, many appear to argue that the cause, and therefore presumably the cure, for the condition of the Third World is due essentially to a single factor. Since there is clearly no agreement on that factor, then perhaps another perspective would be that the issue *is* really complex and involves various elements perhaps differing in importance through time and among countries.

Some have taken their lack of national development as an implication of lessened national prestige, and thereby compensate by alleging that their culture has spiritual and ethical values not found, or lost, in the higher technological societies. Unquestionably, all cultures have some merit; but in the cultural competition that is occurring, and that has always occurred since cultures came into contact with each other, the trend toward a highly urbanized, educated, and industrial–commercial society is relentless and cannot be denied.

In the national quest for development the traditional indigenous cultures virtually everywhere are at a serious disadvantage. Many in these cultures (and some from the outside) wish to preserve some or all of their ancient traditions in the face of this cultural onslaught. Since development requires fundamental changes in society and the economy, these more traditional cultures in reality face the options of adjusting in some important degree or paying the price of slower change – less economic growth, lowered material well-being, and falling ever further behind the more technically-oriented cultures. The choice should be theirs to make but the consequences must be weighed with extreme caution.

There is a great diversity of approaches to development in the form of economic systems operating in the Third World. These range from capitalist/private-enterprise economies in which there is relatively minor government intervention in the operation of the economy, through varying degrees of state intervention and ownership, to the near-total elimination of private enterprises and virtually full governmental control of the economy seen in socialist/communist regimes. On the basis of recent economic growth no one economic system appears to have a clear advantage over the others.

Political systems also exhibit a great range within the Third World. There are relatively open multiparty democracies, semi-democracies or partly free societies, and authoritarian and totalitarian regimes. Again, no one political system appears to provide a greater benefit in terms of economic growth than the others. The only clear association that can be made between economic and political systems is that marxist-based economies (of diverse types) permit no political competitiveness and only very restricted civil liberties.

The diverse and controversial solutions or approaches to development that have been advocated have made it clear that there is little overall consensus and the evidence indicates there is no sure "solution." Probably no theory, and few concepts, with regard to the development process have been accepted by the majority of specialists. Furthermore, the situation in many parts of the Third World is desperate and it is not getting better. The time for resolution is not endless.

It is to be hoped that the following chapters will provide insights into the complex nature of the problem of development. No simplistic solutions have been put forward. Differing options are available and contrasting perspectives are presented. No attempt at conformity or unity of approach has been attempted in these presentations because in the real world these do not exist unless imposed by authorities on a country. We have tried to present as balanced an approach as possible. The reader may assess, interpret, and judge from the writings of these specialists, and so choose the perspective he or she prefers and identify the possible solutions.

J. Norwine and A. Gonzalez

References

Boulding, K. 1980. *Science: our common heritage*. Presidential address, annual meeting of the American Association for the Advancement of Science. San Francisco, California.

Drieske, N. 1986. *Macondo offers Americans portrait of a different South America*. News release, Facets Multimedia, Inc., 1517 West Fullerton Avenue, Chicago, Illinois.

García Márquez, G. 1971. *One hundred years of solitude*. New York: Avon.

Krisnamurti, J. 1970. *Think on these things*. New York: Perennial Library, Harper & Row.

Naipaul, Shiva. 1985. A thousand million invisible men: the myth of the Third World. *The Spectator*, May 18, 1985. 9–11.

Paz, Octavio. 1984. *The labyrinth of solitude: life and thought in Mexico*. New York: Grove Press.

PART I

Why "Third World"?

Approaches to Global Stratification

Abu Hurayra told that God's messenger kissed al-Hasan Ibn Ali, and al-Aqra Ibn Habis who was with him said, "I have ten children and have never kissed one of them." God's messenger looked at him and said, "He who does not show tenderness will not be shown tenderness."

Forty Traditions of An-Nawawi
(*Hadith selection, d. 1277*)

This introductory part provides the basic foundation of terminology and the various approaches and methods of classification in the process of development in the Third World.

Yi-Fu Tuan offers an overview of what constitutes "the good life." There are marked cultural differences in value systems that make a universal definition impossible. There are problems with regard to material prosperity and communal life, and urbanization is fashioning major changes. This author, like others in this book, questions the rewards of modernization.

Allen Merriam examines the term and concept "Third World" and discusses the connotations and the use of other terms, e.g., "East–West," "North–South," and "nonaligned." He concludes that perhaps the most acceptable term is "Third World."

Richard Estes analyzes the background research and four approaches plus his own to formulate a quality-of-life index. He evaluates measures to assess material and international social development and also some of the more subjective approaches (measurement of "happiness" and "satisfaction"). He provides his Index of Social Progress which is also an index of national adequacy of social welfare services.

Alfonso Gonzalez also discusses some of the problems of measuring socioeconomic development and considers the major indexes in use. He presents his own Socio-Economic Development Index and the levels of development, and compares the various indexes in evaluating the world's 58 major countries.

A. Gonzalez

1 *On the rewarding human life*

YI-FU TUAN

Every human being wants the good life. What does the good life mean? The idea of the good or rewarding life no doubt varies recognizably from culture to culture; within a culture that has a complex structure it also varies from social class to social class, and perhaps even from individual to individual. Yet all human beings have certain basic needs, one of which is the means of subsistence. We have to eat. The good life is inconceivable unless the basic bodily needs are met. In China, people greet each other with the question, "Have you eaten?" How can friends enjoy the uniquely human pleasure of conversation or play a game of chess under the gingko tree without food in the stomach? In the English-speaking world, people ask, "How are you?" and the proper response is the reassuring "I am well." Unless one is well, sound of health, questions of the good life will seem beside the point. In French and German, and in English as well, one asks variant forms of the question, "How goes it?" Here life is considered to be a process that moves forward – which, however, is easily derailed by the ill winds of fortune.

Material prosperity

People love life. Closely tied to this love of life is the desire for material prosperity. The Zuni Indians of the American Southwest value corn and other material things because these support life – they make it possible to have many happy and healthy children. In an environment of uncertain rainfall, the Zuni know that the good things they want cannot be obtained without effort. Hence effort is valued. A hard worker is a good Zuni. The good life is one in which hard work is rewarded. When asked to depict the good life, the Zuni draw pictures of abundance such as corn fields and grazing sheep. Their neighbors, the Navajo Indians, depict the good life as "a vision of green and summery landscape able to support its animal and human life."[1]

Is the vision so different in a large and complex agricultural civilization? It is not. Consider China. The good life there is traditionally represented by the three characters *lu*, *fu*, and *shou*. *Lu* means emolument, prosperity, and happiness. *Fu* means prosperity and luck. *Shou* means longevity. To enjoy a long and happy life, one must have the necessary material means. Hard work may earn one the required material compensations: thus by laboring in the field the farmer produces good crops, and through diligence the official gains preferment – a good salary. But, as the Chinese and other peoples throughout the world have realized, hard work never guarantees success. Hence the importance of luck or good fortune. With hard work and good luck, prosperity is assured; and with prosperity one has the wherewithal for living the good life.

In the West, the association of wellbeing with material prosperity has deep roots, reaching back all the way to ancient Mesopotamia. The Sumerians, like all people, cherished life but they did so with a special tenacity because of their convictions that after death the shadowy spirit descended into a dark nether world, where life was at best a dismal reflection of life on earth. What did the Sumerians prize most highly in this world? What made living in it so worthwhile? Answer: "wealth and possessions, rich harvests, well-stocked granaries, folds and stalls filled with cattle large and small, successful hunting on the plain and good fishing in the sea."[2] From the viewpoint of trying to understand the basic human needs this list is significant – indeed poignant – precisely because it is so ordinary. People have always desired successful hunting and good fishing, well-stocked granaries and well-filled stalls and folds. Consider the example of the ancient Greeks. The landscape they loved above all was the deeply humanized, fertile landscape of farms and orchards, not the wilderness – not even the "wine-dark" sea, which the Greeks (for all their dependence on it) tended to view as a stepmother to human beings.[3] Closer to our time is the vision of "the norm of life" in Shakespeare's world. What is it that human beings really want? Shakespeare's answer, echoed in his late plays, is an image of order, peace, honor, and beauty, made possible by the fulfillment of certain material conditions. People did not aspire to be saints or heroes. The strenuous and sacrificial life was a vocation for the few. What most people hoped for was something far more elementary, and expressed by Ferdinand in *The Tempest* as "quiet days, fair issue, and long life." This hope is reiterated by Juno in Prospero's pageant: "Honour, riches, marriage blessing,/Long continuance and increasing."[4]

Delight in the ordinary

Much that is rewarding in life lies in the realm of the norm or the ordinary. Each culture has its own distinctive versions of the norm. When we try to envisage the rewards of life, we need to move beyond abstract concepts to the details of living.

What, for instance, are the specific kinds of experience that give flavor to the nomadic life? What significative images come to mind? Perhaps the following: the open plains and their air of freedom; herders and their flock moving slowly across boundless space; a horseman kicking up dust and streaming like an arrow over the steppe; a family having dinner in the yurt, with the grandmother fondling a sleepy child, and the sound of howling wind and wolves outside.[5] What are the rewards of farm life? A different set of images comes to mind. Rather than pictures of movement and openness, we see those that evoke attachment to place, intimacy with soil and the seasons of nature, continuity and stability, rootedness, human warmth generated by sowing and harvesting together, the feeling of contentment and peace after periods of hard work. The goodness of country life is most convincing when its timeless sensuous qualities are stressed. Passages from Hesiod's *Works and days* capture the sound of chirping crickets; the heat of summer; the taste of fresh goat's milk and of wine; the wonderful sense of satiety after a good meal, when one sits under a tree and is cooled by a gentle breeze. These impressions, recorded around the 7th century BC, are just as real and appealing today.

Farmers sell their agricultural surplus and the things they manufacture at home. They are also traders. What is happiness – what can give a greater sense of being alive and vital among relatives, friends, and strangers – than buying and selling in a busy African market? Here is a picture of a market day among the Bangwa of Cameroon. People spill down the narrow paths leading to the open space before the chief's big house "from the earliest morning hour, loaded with raw and cooked food, wine, oil, woven mats, goats. Women set up countless palm-wine bars and Ibo traders arrive from yesterday's market in a neighboring village bringing their cloths and 'fancy goods.' Purchases are quickly made and the rest of the time is spent greeting friends, exchanging kola nuts, or drinking a cup of wine with a relative or an acquaintance."[6] At times the exuberance makes the numerous activities of the marketplace seem a wildly gay chaos. Clearly the Bangwa enjoy their markets. Indeed to the chiefs, nervous of the innocence of their young wives, they enjoy the markets too much. As night falls, a royal servant might appear beating a gong to send people back to their homes and farms.

City: cosmic and modern

The ancient Greeks uttered some extremely pessimistic sayings. "Call no man happy till he dies; he is at best fortunate" (Solon). "Best it is not to be born; second best is to die young" (Sophocles). And not only philosophers, but ordinary people as well, raised the question, "Why is it better to be born than not to be born?" The Greeks, by these expressions, recognized the harshness of life. To be born is to be born into suffering. There may be the natural vitality of youth but that vitality cannot last. Life, moreover, is uncertain. Human beings are the playthings of moody gods. What the ancient Greeks had said, other peoples – even if they did not say so – must have also felt. A common human dream is therefore of a nature that is totally nurturing – the Isles of the Blessed, the Garden of Eden, or the Taoist paradise. Another dream, antipodal to luxuriant organic nature, is of cosmos. Life would be more ordered, more certain, if something of the regularity of the stars could be brought down to earth. The result is the geometrical, mineralized, cosmic city.

The city appeals to the human spirit, especially its vaulting ambitions. It is not ordinary either as architecture or as a social world. The higher the aspiration, the longer, it is true, the fall; and the story of the city is not only that of human achievement but also that of forced labor, crime, extravagant wealth, grotesque luxury, depravity, and the abuse of power. Nevertheless the achievement is there. Monumental walls, temples, and palaces – considered simply as built forms and physical presences – have the power to expand one's vision of the possible far beyond that which can be envisaged in a farmstead or village.[7] The city has, of course, radically altered its physical appearance in the course of time. Perhaps the most dramatic of the physical changes in the long history of the city is the introduction of adequate lighting. Day conquers night, light overcomes darkness. If illumination were a human desire, it could be put into practice only on a very limited scale until the appearance first of gas light, in the early part of the 19th century, and later of electric light. Before these technologies were invented, the city – no matter how splendid and proud in the day – was plunged into darkness, just like the countryside, after sunset. But light has symbolic meanings to humankind as well. Illumination is not only of streets and shops for a practical purpose, it is also of the mind. After dark, human consciousness, instead of being allowed to dim naturally, is stimulated to an even more intense level as people attend a learned lecture, the theater, or a symphony concert. The city at night beckons. In the dark countryside, it appears in the distance as a glowing dome on the horizon. Light signifies life. The illuminated city hints at more life, more opportunities for work, greater stimulation of the senses, more life of the mind.[8]

Communal ideals

In folk cultures and Third World countries, a deep source of human satisfaction is communal life. For close human ties to exist, the community must be small and share a common unexamined tradition. Moreover, there must also be a sense of a threatening "other," whether this other is harsh undependable weather or alien human groups. A purpose of technology is to remove the threats of nature. To the extent that the threat is removed, the human bonds that are knit in response to it necessarily also weaken. A purpose of social reform is to remove conflict between groups, and to the extent that it succeeds internal cohesion within the group weakens. Thus progress on both the technological and social fronts has the effect of cooling the warmth of traditional communal life, including not only love but also the destructive passions of envy, jealousy, and hatred.

A third source of corrosion is increasing reflectiveness on the part of individuals. This too is a mark of progress – intellectual progress. As people become more affluent and are more conscious of their own personal values, they will want to build their own house on a spacious ground, which has the consequence of isolating them from other people living in similar houses. Likewise, as people become more critical of communal values and religious–ethical beliefs, they will be inclined to explore the possibility of building their own mental edifice tailored to their own knowledge and experience. They will live in this mental edifice as they live in the large physical house, and will come to feel isolated from their neighbors ensconced in mental edifices similarly elaborated to suit their own uniquely realized perceptions and knowledge. What if we are not systematic thinkers but poets? The same isolation and loneliness result. Precise, fresh, and evocative language is not easy to understand – perhaps no easier than the language of systematic thought. Poetic language can communicate only if people take the time to listen with focused attention. But this is an acquired skill, difficult to practice and rarely practiced even among the well-educated. Poets themselves can form a community only if they speak to each other (as we all do when we genuinely wish to belong) in clichés. Lovers will feel estranged if they address each other with the eloquence of Romeo and Juliet; so, to maintain the bond, they instinctively reduce their intellect to that of babies.

Modern society, with its vast bureaucracy and impersonality, is alienating. It is indeed, but not only for the above reasons: it is also alienating because it promotes those qualities which people of the First World – at least in the abstract – admire: namely, independence of mind, a critical outlook, the ability to build one's own moral–intellectual edifice, and the courage to recognize one's unwanted uniqueness – hence one's separateness from others.

People of the Third World may well wonder about the rewards of modernization. Is it possible to illuminate the streets without also illuminating the mind? Can one have a good life if one keeps on thinking about it, thus distancing the self from the direct, unreflective experience of nature and of other people? "The good life is a life in which people never cease to pursue the question, what is the good life?" This is a boast of thoughtful modern man and woman. Does it ring hollow?

Notes

1 E. Z. Vogt & E. M. Albert 1966. *People of Rimrock: A study of values in five cultures*, 42, 283. Cambridge: Harvard University Press.
2 S. N. Kramer 1963. *The Sumerians*, 262–3. Chicago: University of Chicago Press.
3 W. R. Paton (transl.) 1917. *The Greek Anthology*. vol. 3, 15. New York: Putnam.

4 For the idea of the norm of life in Shakespeare, see L. Trilling 1972, *Sincerity and authenticity*, 39. Cambridge: Harvard University Press.

5 *Altan tobci* (nova), Scripta Mongolica edition 1953, vol. 2, 55–7. Cambridge: Harvard University Press. S. Jagchid & P. Hyer 1974. *Mongolia's culture and society*, 111. Boulder, Col.: Westview Press.

6 R. Brain 1976. *Friends and lovers*, 154. New York: Basic Books.

7 L. Mumford 1961. *The city in history*, 31. New York: Harcourt, Brace & World.

8 I have developed this idea in "The city: its distance from nature," *The Geographical Review*, vol. 68, no. 1, 1978. 1–12. See also M. Bouman 1984: "City lights and city life: a study of technology and urbanity." (Ph.D. dissertation, University of Minnesota.)

2 *What does "Third World" Mean?*

ALLEN H. MERRIAM

The term "Third World" represents one of the significant additions to the vocabulary of the 20th century. The 1985–6 edition of *Books in print* lists more than 60 English-language books with the term in the beginning of their title. "Third World" refers to the nations of Asia, Africa, and Latin America, which generally are characterized by relatively low per capita incomes, high rates of illiteracy, agriculturally-based economies, short life expectancies, low degrees of social mobility, strong attachments to tradition and, usually, a history of colonization (Sachs 1976, Mountjoy 1979, Hoogvelt 1982, Guernier 1982). These countries are thus distinguished from the First World (and Western industrialized democracies and Japan, with Israel and South Africa sometimes included) and the Second World (the Communist bloc of European nations of the Soviet Union, Poland, East Germany, Hungary, Romania, Bulgaria, and Czechoslovakia).

As with many systems of classification, however, this one may suffer from the dangers of oversimplification. What does it mean to divide the world into thirds? What are the biases involved in assigning nations to third place? Are there any preferable terms for describing the various segments of humanity?

The importance of language choice and labels in international affairs is readily apparent. Some years ago the International Red Cross requested permission to investigate allegations of mistreatment of prisoners of war captured during the civil war in Rhodesia (now Zimbabwe). The Salisbury (now Harare) government responded that the captives were not "prisoners of war" but instead were "guerrillas" and "terrorists" and therefore not covered by the Geneva Convention. How the government named the soldiers determined its treatment of them.

Semanticists (Hayakawa 1972) normally speak of two levels of meaning in language: *denotative* meanings which refer to the public, dictionary definitions of words, and *connotative* meanings which include all the mental associations, feelings, and private assumptions triggered in a listener's mind. The word "Zionism," for example, may denote a specific ideology of Jewish nationalism, but the term possesses vastly different connotations to Israeli Jews and Palestinian Arabs. Given the potential power of words in international relations, it is appropriate to consider the semantic implications of the term "Third World."

The concept of a "Third World" is European in origin. Safire (1978) traced the phrase to France in the 1940s as a description of the political parties distinct from both de Gaulle's *Rassemblement des Peuple Français* (RPF) and the Fourth Republic, while Clegern (1978) compared the idea to the Third Estate, the rising but under-represented bourgeoisie in the French Revolution of 1789. Wolf-Phillips (1979) credited the French demographer, Alfred Sauvy, with coining the term in 1952. Since the early 1960s the phrase has gained increasing acceptance as a positive concept signifying the new and experimental arena of global politics bound to neither Western capitalism nor Soviet socialism.

The most obvious observation to be made about the term is that "third" is inferior to "first" or "second." The phrase therefore might be viewed as offensive to the peoples so labelled. This indignation could be especially acute for the Chinese, who traditionally described their nation as *chung kuo* (Middle Kingdom), reflecting an assumption that China formed the center of the universe. Illustrative of the historic Chinese view of their superiority was the message Ch'ing Dynasty emperor Ch'ien Lung (reigned 1736–95) sent to King George III of England: "Swaying the wide world . . . our dynasty's majestic virtue has penetrated into every country under Heaven, and Kings of all nations have offered their costly tribute by land and sea . . . We possess all things. I set no value on objects strange and ingenious, and have no use for your country's manufactures." (Lum 1973). Certainly he would not have appreciated being relegated to third place.

Further bias may be observed in the criteria used in dividing the world into thirds. Clearly, a mixture of economic, political, technological, and social factors is involved. But to define development in terms of wealth and technology shows a decidedly Western bias. The very notion of "development," in fact, presupposes a commitment to materialism, change, progress, and science, which are basically Western values (Stewart 1972). In some ways the "poor" cultures of the Third World are rich psychologically and spiritually, enjoying a contentment and sense of tradition sorely lacking in hectic, ulcer-ridden, depersonalised industrial societies. To many Buddhists, for example, inner peace is more valuable than a high Gross National Product. The highest divorce and suicide rates occur in the First and Second Worlds. If personal happiness were our criterion, the Third World might rank first.

The Third World would again be first if our standard of judgment were the chronology of the human species. Leakey and others (1982) have traced the earliest known forms of human life to Africa. Cognizant of this, the editors of a journal of black thought deliberately chose *First World* as their publication's name, arguing that as descendents of Africa "we had a valid claim to call ourselves people of the first – or original – world . . . we have the obligation to reject definitions imposed upon us by another people who have seen us as subjects for degradation and objects for exploitation." (*First World* 1977).

Or, if population were the criterion, the Third World would again be first. Latest estimates indicate that approximately 75% of the earth's people live in the Third World, with only about 18% in the First World, and the remaining 7% in the Soviet-led Second World. While some might argue that China and her Communist neighbors of North Korea, Laos, Kampuchea, and Vietnam belong in the marxist or Second World, such placement seems inappropriate for two reasons. First, Sino-Soviet relations have become sufficiently strained that these two powers cannot be viewed as a united force in world affairs. Secondly, China's leaders have expressly stated their identification with the Third World: in welcoming former Philippines President Marcos to Beijing in 1975, Vice-Premier Deng Xiaoping declared "China is a developing socialist country belonging to the third world." (*Peking Review* 1975, Larken 1975). Recent experiments with capitalism, and the dramatic *People's Daily* editorial of December 7, 1984 ("China calls . . . " 1984) claiming that marxism cannot solve all of today's problems, further suggest that China belongs under the Third World umbrella (Harris & Worden 1986).

It is instructive to note that near the end of his life Mao Ze-dong formulated an interesting if unorthodox theory of three worlds (*Chairman Mao's* . . . 1977, Dai 1978). Mao viewed the First World as consisting of the superpowers (Soviet Union and United States) whose imperialistic policies, he felt, posed the greatest threat to world peace. Mao placed the middle powers (Japan, Canada, and Europe) in the Second World. Africa, Latin America, and Asia (including China) formed the Third World, the peoples whom Mao believed to be the best hope for revolutionary struggle.

In recent years critics have argued that the concept of three worlds is no longer adequate and that we must now more realistically speak of four worlds. Manuel & Posluns (1974) claim that the Third World does include the emerging nation states which formerly were colonies, but that a fourth world also exists consisting of the earth's aboriginal peoples. They cite the Indians of the Americas, the Lapps of northern Scandinavia, and Australia's aborigines as groups which remain powerless amid more dominant cultures, and are therefore still colonized.

Working from an economic rather than political standard, *Newsweek* magazine ("To have . . . " 1975) also asserted the existence of four worlds. According to this scheme the Third World includes developing nations with significant economic potential, such as Argentina, Egypt, Indonesia, Iran, Mexico, Nigeria, South Korea, Saudi Arabia, and Taiwan, whereas the Fourth World would designate "the worst economic hardship cases." In the latter category *Newsweek* placed such countries as Bangladesh, Chad, Haiti, India, Pakistan, Togo, and Zaire.

Not to be outdone, *Time* magazine subsequently suggested ("Poor . . . " 1975) that we must even add a fifth world to our nomenclature, an opinion shared by the American diplomat George Ball (1976). Under this plan the Third World would include those nations which, although still "undeveloped," possess important natural resources. Brazil, Mexico, Malaysia, Saudi Arabia, Iran, Nigeria, and Venezuela were placed in this category. The Fourth World was portrayed as countries whose economic development depends largely on industrial powers for the export of goods and the import of technology. Egypt, India, Liberia, Peru, Tanzania, and Thailand fit this description. *Time* reserved the Fifth World for "the globe's true basket cases," meaning impoverished countries such as Afghanistan, Bangladesh, Chad, Ethiopia, Nepal, and Somalia.

Having identified some of the possible negative connotations of the term "Third World," we might turn to an analysis of alternatives. Some scholars (Haas 1956, Northrup 1946) have made meaningful distinctions between "the East" and "the West." Typically this dichotomy emphasizes the subjective, intuitive, unity-with-nature, group orientation of Eastern cultures in contrast to the individualistic, objective, rationalist, and scientific tendencies of the West. Even India's great humanitarian poet Rabindranath Tagore accepted this basic division, claiming that two literary epics – Shakespeare's *The Tempest* and Kalidasa's *Shakuntala* – symbolized the opposing attitudes toward man and nature. Tagore felt (1916, 1961; Merriam 1974) that humanity's true psychic wholeness requires a synthesis of the East's spiritual joy with the West's scientific law. But while the East–West distinction may help our understanding of the divergences in Oriental and Occidental philosophical systems, and even this assumption has been challenged (Rosan 1962), the terms blur important political differences found within the West, such as between East Germany and West Germany, and they ignore Africa altogether. Thus, the East–West division is not a workable alternative to "Third World."

Another two-part designation divides the world into "North–South," reflecting the fact that Northern Hemisphere nations tend to be more economically advanced than their southern counterparts. But this dichotomy fails to distinguish between northern countries as diverse as the United States and the Soviet Union, and contains distortions resulting from the location of South Asia, two-thirds of Africa, and all of Central America in the northern hemisphere. Therefore, "North–South" is of doubtful usefulness.

Another point of view holds that humanity is essentially one, and that our verbal descriptions should not fragment it into competing and hostile groups. One advocate of such universalism was the 17th-century Czech educator Comenius, who affirmed: "We are all citizens of one world, we are all of one blood . . . Let us have but one end in view, the welfare of humanity." (Fersh 1974). Sharing this sentiment was the 19th-century Persian mystic Baha'u'llah, whose disciple Abdul'-Ba'ha declared: "All are servants of God and members of one human family. God has created all and all are His children . . . we are all the waves of one sea." (*Baha'i World Faith* 1976). The notion of one world was echoed more recently by Tanzania's former president Julius Nyerere (1970) when he told the United Nations: "We believe that all mankind is one, that the physiological differences between us are unimportant in comparison with our common humanity." And India's late prime minister, Indira Gandhi, specifically objected to the term "Third World" since it appears to deny the oneness of humanity ("Mrs. Gandhi . . . " 1973). Said Mrs. Gandhi: "We are responsible not to individual countries alone, but to the peace and prosperity of the whole world." But although world unity represents an admirable philosophical ideal, the concept is so comprehensive and even Utopian that it makes impossible the political and economic distinctions upon which governments, multinational corporations, and social scientists base decisions daily. Thus, it is not a viable alternative either.

One of the more creative words proposed to describe the Third World is "Bandungia," used by Vera Micheles Dean (1957). This word takes its name from the city in Indonesia which hosted the important meeting of Afro-Asian nations in 1955. But while the Bandung Conference marked a significant symbolic step in the growth of post-colonial assertiveness by Third World nations, the term "Bandungia" has not caught on, and today sounds more like a place in science fiction than a description of global geopolitics.

Another appellation often heard is "nonaligned nations." This phrase sometimes is used synonymously with "Third World," and the two are closely related both historically and ideologically. Many of the participants at the Bandung Conference sought to institutionalize their movement, and heads of state of 25 nonaligned countries convened in Belgrade in 1961. Subsequent assemblies – in Cairo in 1964, Lusaka in 1970, Algiers in 1973, Colombo in 1976, Havana in 1979, New Delhi in 1983, and Harare in 1986 – have witnessed a steady and impressive growth. The term "nonaligned," however, has some drawbacks, not the least of which is its inaccuracy. Such members as Cuba and Nicaragua are, at least from the First World's view, closely aligned with the Communist camp. For many years the nonaligned membership list omitted Pakistan and Turkey due to their military ties to the West, through CENTO and NATO respectively, yet both countries are clearly part of the Third World. And the fact that nonaligned countries are aligned with each other makes their name misleading if not contradictory.

Several sets of initials are sometimes encountered. "LDCs" (Less Developed Countries) and "NICs" (Newly Industrialized Countries) have the advantage of brevity but suffer from the narrow perspective of defining development in solely economic terms. The word "less" also carries negative connotations, and it seems awkward to apply the name "newly" to countries with ancient civilizations. A related phrase which may be among the most acceptable alternatives is "developing nations," but even that has limitations. The marxist writer Fred Carrier (1976) has objected to it because it seems to suggest that Third World nations will merely follow First World patterns and develop along the "same capitalist path." Carrier prefers the term "Third World." Moreover, "developing nations" seems too inclusive to apply only to Third World societies. After all, the United States is developing; Japan and the Soviet Union are evolving. Indeed, the devotion to technological advancement and social change in Western nations means that they may be developing faster than Third World countries.

Yet another term sometimes employed is "Group of 77." This name refers to the 77 nations which, at the first meeting of the United Nations Conference on Trade and Development (UNCTAD) in Geneva in 1964, formed a separate entity to press for bigger and more vigorous international development programs. But "Group of 77" is a cumbersome label, and it is inaccurate as more than a hundred nations are now members. Several additional terms, used at various times historically but now deemed unacceptable due to blatant assumptions of cultural arrogance, include "undeveloped areas," "backward nations," and "primitive societies."

So, is there one world? Or two? Or three? Or five? Or five billion? – for theoretically one could argue that each human being perceives and interprets reality in a unique way, and therefore there are as many "worlds" as people. My conclusion is that we should continue to use the term "Third World" despite its possible deficiencies. To paraphrase Winston Churchill, it is the worst term we have, except for all the others. No other phrase seems to achieve greater clarity and simplicity.

And the term does have some advantages. Three is a convenient and workable number large enough to make useful distinctions between the world's major ideological groups, although this assertion may reflect a typically American penchant for threes (Condon & Yousef 1975). Easily translated into other languages such as Spanish (*el tercer mundo*) and French (*le tiers monde*), the concept of "Third World" can allow scholars and government planners to describe, analyze, and predict political, economic, and social commonalities among nations as culturally diverse as Brazil, Burundi, and Burma. Moreover, the phrase has proven rhetorically effective to leaders within the Third World itself. The term forms a rallying point, a persuasive slogan, a source of identity for the peoples so labelled, as evidenced by calls for Third World solidarity at the United Nations, creation of a Third World Development Bank, and a restructuring of the globe's news and media systems within the New World Information Order (NWIO) espoused by UNESCO (Stevenson & Shaw 1984, McPhail 1981, MacBride 1980).

Thus, despite its imperfections, the term "Third World" can be viewed as a meaningful addition to our vocabulary. It is manageable, functional, and forceful. Whether in the speeches of Fidel Castro, the poetry of Aime Cesaire, or the economic analyses of Mahub ul-Haq, the term clearly has come to represent an ideology of its own. More than merely a socio-economic designation, it connotes a psychological condition, a state of mind encompassing the hopes and aspirations of three-fourths of humanity.

References

Baha'i World Faith 1976. Wilmette, Ill.: Baha'i Publishing Trust.

Ball, G. 1976. *Diplomacy for a crowded world.* Boston: Little, Brown:

Books in print 1985. New York: R. R. Bowker.

Carrier, F. J. 1976. *The Third World revolution.* Amsterdam: Gruner.

Chairman Mao's theory of the differentiation of the three worlds is a major contribution to Marxism–Leninism 1977. Beijing: Foreign Languages Press.

China calls rigid adherence to Marxism "stupid" 1984. *The New York Times* (December 9) I, 21.

Clegern, W. M. 1978. What is the Third World? Paper presented to 2nd Nat. Conf. on the Third World, Omaha, Nebraska.

Condon, J. C. & F. Yousef 1975. *An introduction to intercultural communication.* Indianapolis: Bobbs-Merrill.

Dai, S. Y. 1978. The Third World and Mao Tse-tung's three world theory/strategy. Paper presented to 2nd Nat. Conf. on the Third World, Omaha, Nebraska.

Dean, V. M. 1957. *The nature of the non-Western world.* New York: Mentor.

Fersh, S. 1974. *Learning about peoples and cultures.* Evanston, Ill.: McDougal, Little.

First World: an International Journal of Black Thought 1977. **1**, 1.

Guernier, M. 1982. *The Third World: three-quarters of the world.* Elmsford, NY: Pergamon Press.

Haas, W. 1956. *The destiny of the mind: East and West.* London: Macmillan.

Harris, L. C. & R. L. Worden (eds.) 1986. *China and the Third World: champion or challenger?* Dover, Mass.: Auburn House.

Hayakawa, S. I. 1972. *Language in thought and action*, 3rd edn. New York: Harcourt Brace Jovanovich.

Hoogvelt, A. 1982. *Third World in global development.* London: Macmillan.

Larken, B. 1975. China and the Third World. *Current History* **69**, 75–9.

Leakey, R. E. & R. Lewin 1982. *Origins.* New York: E. P. Dutton.

Lum, P. 1973. *Six centuries in East Asia.* New York: S. G. Phillips.

MacBride, S. (ed.) 1980. *Many voices, one world.* Paris: United Nations Educational, Scientific and Cultural Organization.

Manuel, G. & M. Posluns 1974. *The Fourth World: an Indian reality.* New York: Free Press.

McPhail, T. L. 1981. *Electronic colonialism; the future of international broadcasting and communication.* Beverly Hills, Cal.: Sage.

Merriam, A. 1974. Rabindranath Tagore's concept of man. *Indian Horizons* **23**, 21–32.

Mountjoy, A. B. (ed.) 1979. *The Third World: problems and perspectives.* New York: St. Martin.

Mrs. Gandhi welcomes dawn of era of detente 1973. *India News* (September 14), 2.

Northrop, F. S. C. 1946. *The meeting of East and West.* New York: Macmillan.

Nyerere, J. K. 1970. All mankind is one. In *Sow the wind, reap the whirlwind: Heads of State address the United Nations*, M. H. Prosser (ed.), II, 824–5. New York: Morrow.

Peking Review 1975. (June 13), 9.

Poor vs. rich: a new global conflict 1975. *Time* (December 22), 34–42.

Rosan, L. J. 1962. Are comparisons between East and West fruitful for comparative philosophy? *Philosophy East and West* **11**, 239–43.

Sachs, I. 1976. *The discovery of the Third World*, trans. M. Fineberg. Cambridge, Mass.: MIT Press.

Safire, W. 1978. *Safire's political dictionary: the new language of politics.* New York: Random House.

Stevenson, R. L. & D. L. Shaw 1984. *Foreign news and the new world information order.* Ames: Iowa State University Press.
Stewart, E. C. 1972. *American cultural patterns: a cross-cultural perspective.* Pittsburgh: Regional Council for International Education.
Tagore, R. 1916. *Sadhana: the realization of life.* New York: Macmillan.
Tagore, R. 1961. *Towards universal man.* New York: Asia Publishing House.
To have and have not 1975. *Newsweek* (September 15), 37–45.
Wolf-Phillips, L. 1979. Why Third World? *Third World Quarterly* **1**, 105–14.

3 *Toward a "quality-of-life" index: empirical approaches to assessing human welfare internationally*

RICHARD J. ESTES

The need for a "quality-of-life" index

Since the mid-1940s social scientists have undertaken a number of significant efforts toward developing research tools that could effectively assess the capacity of nations to provide for the basic needs of their people. These efforts have examined both the capacity of nations to provide for basic human needs at discrete points in time and also their changing capacity vis-à-vis the adequacy of their social provision over time. Referred to as research on "level of living," "social wellbeing," "quality of life," "physical quality of life," "social welfare," and "human welfare," etc. these efforts have attempted to measure a more inclusive range of social phenomena than is possible when only economic factors are considered. In general, each approach to "human welfare" assessment has sought to combine various indicators of social wellbeing into composite indexes that could be used for purposes of cross-national comparative analyses. Within individual nations, for example, these approaches to research have been used to assess the changing social needs of discrete population groups (e.g., of men versus women, of children versus older people, and of minorities).

An extensive body of cross-national comparative research currently exists that describes the conceptual, methodological, and even political issues involved in constructing social indicators for purposes of international research (for a review of this literature see Estes 1983). The majority of these studies, however, are impaired by serious conceptual problems; by problems of missing, incomplete, or otherwise unavailable data; or by the paucity of researchers and statisticians trained to undertake such research. National and international political tensions have contributed to the difficulties of data collection, too, as have the not entirely unreasonable concerns of governments regarding the potential misuses to which politically sensitive human-welfare data might be put. Consequently, few international comparative studies are available that use the world as the unit of analysis, and often those that *are* to be found are more statistical than analytical in nature.

Because of the problems incurred in carrying out large-scale comparative studies, an urgent priority before the international social-research community is the development of new analytical tools for use in assessing the extent of national and international success in meeting the basic needs of the world's growing population. These tools are needed to provide social development ministers, planners, service directors, and researchers with a continuing source of reliable information for use in formulating realistic development objectives. Analytical tools are needed also to establish rational priorities among the competing alternatives for international development assistance. They are needed, too, to provide greater accountability among development-assistance organizations for the substantial expenditures being made by them to promote social and economic development objectives internationally.

Assessing national and international social development

At least five closely related international research approaches have been undertaken in an effort to construct the analytical tools needed by specialists in national and international social development. A sixth approach, which focuses on subjective aspects of wellbeing, attempts to substitute the psychological experience of individuals for the more abstract statistical patterns.

GROSS NATIONAL PRODUCT (GNP)

The most successful approach to assessing change in patterns of national and global development has been Gross National Product (GNP). As an analytical tool, GNP measures the monetary value of all the goods and services produced by a nation at discrete points in time. Developed during the 1930s and 1940s in response to widespread global economic recession, unemployment, and underproductivity (Morris 1979), GNP has served economists and political leaders as an effective and reliable tool for measuring changes in national and international economic trends.

As a measure of "human welfare" or "social wellbeing," however, GNP has never been satisfactory. GNP measures only those economic activities to which a discrete monetary value (a "price tag") can be attached; and it was not designed to take into account the subjective valuations that people place on critical noneconomic experiences. As a specialized social-science tool proven to be useful in economic analyses, forecasting, and planning, GNP simply cannot incorporate all of the diverse *social* phenomena required to assess changes in "human welfare" over time. Indeed, efforts to equate GNP with human welfare, or even to use GNP as an indicator of the differential levels of human welfare that characterize nations at various points in their development, have been sharply criticized (Drenowski 1974, Morris 1979).

UNITED NATIONS (UN)

With the exception of the pioneering work of economists Joseph Davis (1945) and M. K. Bennett 1951), the most significant work on developing more inclusive approaches to assessing changes in the global social situation was initiated under the auspices of the UN. Acting in conformity with Article 55 of its Charter, the UN established an Expert Group that was charged

> to prepare a report on the most satisfactory methods of defining and measuring standards of living and changes therein in the various countries, having regard to the possibility of international comparison (UN 1954).

After meeting for two years, the Expert Group was unable to produce a definitive response to such an ambitious charge, but it did identify the need to distinguish between the concepts of "standard," "norm," and "level of living." Its report emphasized the need for quantitative measures of human welfare in the areas of health, nutrition, housing, employment and education (UN 1954). The group believed that the development of quantitative "indicators" of human welfare was essential to the establishment of objective standards that could then be applied to analysis and comparison of changes in the "level of living" that would characterize member states over time.

The 1954 report stimulated considerable discussion within the UN Economic and Social Council (ECOSOC) as well as in the General Assembly. An interagency commision was formed to promote cooperation between the various specialized agencies of the UN (e.g., the ILO, WHO, and UNESCO), and a second Expert Group was formed to implement the 1954 recommendations. Their report (UN 1961) advanced the work considerably and listed 12 components to be included in a "level-of-living" index: health, food and nutrition, education (including literacy and skills training), conditions of work, employment situation, aggregate consumption and savings, transportation, housing, clothing, recreation and entertainment, social security, and human freedoms.

The Group also identified discrete quantifiable indicators that could be used to recast each of the components (e.g., expectation of life at birth, infant mortality rate, and crude annual death rate were to serve as indicators of level-of-living in relation to health). The conceptual and data-collection principles formulated by the Group were adopted with few changes and subsequently outlined in handbook format for use by the UN in collecting data from member countries (see UN 1964).

Recognizing the peace-keeping importance of promoting social development more systematically throughout the world, the UN established the

Research Institute on Social Development (UNRISD) in Geneva, Switzerland, in 1961. The purpose of UNRISD then, as now, was to engage in systematic research into the dynamics of social development nationally and in the context of world economic development. The work of UNRISD focused initially on highly technical studies that identified more effective approaches to assessing the level-of-living in a variety of cross-cultural contexts (see Baster & Scott 1969; Drenowski 1970, 1974). One especially important study undertaken by a UNRISD team led by Donald McGranahan examined "the correlates of (international) socioeconomic development" (McGranahan 1972). Using a system of empirically derived statistical "correspondences," the McGranahan team was able to identify particular *economic* factors that most closely predicted a nation's subsequent pattern of *social* development.

Equally ambitious approaches at measuring the nature and dynamics of level of living throughout the world have been undertaken by other units of the UN. The Research Division of the UN Education, Scientific and Cultural Organization (UNESCO) in Paris, for example, has made invaluable contributions to this effort (e.g., UNESCO 1976) as has the UN European Social Development Programme (UNESDP 1976). Even though the research efforts of each of these UN agencies carries its unique theoretical stamp, all share the same fundamental commitment to the development of reliable approaches to an assessment of human welfare that incorporates a broad range of social and economic phenomena. Most are committed to a "basic needs" approach – one that emphasizes the satisfaction of the most fundamental social and material needs of people, irrespective of country of residence or political or economic system (see Streeten *et al.* 1981).

ORGANIZATION FOR ECONOMIC COOPERATION AND DEVELOPMENT (OECD)

A somewhat different approach to assessing adequacy of national and international social provision is that undertaken by the Paris-based Organization for Economic Cooperation and Development (OECD).

The OECD is an intergovernmental organization that brings together 24 of the world's most industrialized nations for the purpose of expanding world trade between both member nations and less developed countries. Altogether, the foreign trade of OECD member countries accounts for about 70% of world commerce. Made up of the world's richest nations, the OECD also carries a "social agenda," one that seeks to promote the social and economic development of less developed nations. In response to this broader social commitment the OECD has developed its own approach to human welfare assessment, one that differs considerably from those developed by the UN and its specialized agencies (see OECD 1977a, 1977b).

In essence, the OECD approach to human-welfare assessment focuses on the concept of "social wellbeing." This concept is used to promote national improvements in two areas: (a) macro-level economic development and (b) advances in micro-level social and economic conditions that enhance the social wellbeing of individual citizens living in OECD nations. National development objectives in the first area have been largely achieved through the establishment of a successful system of "National Accounts", which guides economic planning within and between OECD member states (see OECD 1983, 1984). Work on the second objective – that of assessing and promoting the social wellbeing of individual citizens – is continuing through a series of discrete phases:

Phase I –establish a consensus on a list of social concerns for which social indicators are to be developed;
Phase II – develop a set of social indicators "designed to reveal with validity the level of 'wellbeing' for each social concern . . . and to monitor changes in those levels over time" (OECD 1976, p. 20);
Phase III – make measurements using the social indicators developed in Phases I and II;
Phase IV – link "wellbeing" conditions to practical policy measures and options.

Work on Phases I and II has been largely completed (OECD 1982b) and, in the main, the effort has succeeded in achieving the degree of member-nation commonality and conceptual simplicity that was sought by the original team of investigators (see OECD 1973). The current list of "shared social concerns," for example, identifies 8 areas of "primary" concern and 15 areas of "subconcerns" including health, education and learning, employment and quality of working life, time and leisure, command over goods and services, physical environment, social environment, and personal safety. Further, these concerns have been operationalized through a system of 33 social indicators (see OECD 1982b; p. 13), each of which serves as a focal point for specialized OECD data gathering and analytical work.

For the most part, the list of social concerns identified by the OECD differs only in minor respects from the 12 components of "level-of-living" identified by the earlier UN Expert Groups. The resemblance is particularly striking with respect to the OECD choice of social indicators. However, the OECD shift from exclusively conceptual work to actual data gathering and analysis distinguishes it significantly from the earlier UN groups. Implementation of the OECD data-collection activities marked the implementation of the third phase of its project. Currently, collection and analysis of comparative social wellbeing data are continuing research activities both of the OECD itself and of its member nations (see OECD 1982b).

However, the OECD goal of using the findings obtained from its social analyses to make some favorable impact upon national social and economic policies at the micro-level – i.e., implementation of Phase IV objectives – is far from being realized. Indeed, current economic problems within the majority of OECD nations are of such a magnitude that even the data-gathering activities of Phase III are in jeopardy (Verwayen 1984). Even so, the OECD is continuing its efforts at attempting to influence the social development of member governments so that increasing amounts of fiscal resources will be committed to improving the social wellbeing of individual citizens (see OECD 1981, 1982a, 1985).

Despite the attention that has been generated by the OECD approach, the full significance of its work has yet to be determined, especially because the policy development and implementation phases are only beginning. Even should OECD efforts at that level prove to be successful, its approach to improving the social wellbeing of citizens residing within OECD member nations may contribute relatively little to an improvement in the living conditions of persons who reside in countries that are *not* OECD members, i.e., all of the world's developing and least developing countries. The populations of these countries, which make up 75% of the world's total, face social and economic conditions that differ dramatically from those to be found in economically advanced nations. The social-wellbeing solutions offered by the OECD to its "first world" membership cannot be expected to address adequately the assessment and planning needs of its more impoverished neighbors to the south and east. Consequently, at least in my judgement, current OECD approaches to social-wellbeing assessment, even if successful, are not likely to result in a composite index of social wellbeing that possesses universal applicability. The need for a more universal assessment "yardstick" has been recognized by the OECD itself (OECD 1976), but currently neither the organization nor its members have identified the need as a priority on its research agenda.

PHYSICAL QUALITY OF LIFE INDEX (PQLI)

The social-indexing work of Morris D. Morris and his colleagues at the Washington-based Overseas Development Council (ODC) has received considerable international attention in recent years (Morris 1979). Interest in the Council's work stems from the simplicity of the Morris model for global social assessment, especially in comparison with the more comprehensive models sought by the UN, UNESCO, the OECD, and other international development-assistance organizations.

Essentially, in an effort to minimize the difficult conceptual and methodological problems that blocked the work of other investigators, the Morris team constructed a three-item "Physical Quality-of-Life Index" (PQLI). The PQLI examines variations over time in (a) national rates of infant mortality, (b) years of average life expectation at birth, and (c) rates of adult literacy. These indicators of social development were selected on the basis of (a) their relative theoretical independence from measures of economic development and (b) their "results" orientation. All three indicators focus attention on the social *accomplishments* of nations, rather than on the nature, extent, or amount of resources required to bring about these achievements (e.g., time, money, and human effort).

Morris applied the PQLI to 74 nations at various points in time between 1947 and 1973. His results were intriguing and they provided further evidence for the inappropriateness of GNP as the primary indicator of changes in level of national social development. The correlation of GNP with PQLI, for example, was very weak, thereby confirming that the two indexes measure different phenomena. He also showed that a simplified approach to assessing world social progress can yield results that are both analytically interesting and politically useful in redirecting global development-assistance resources to those nations with the most urgent social needs.

The simplicity of the PQLI model, however, is its major weakness. Of the three indicators used to construct the index, two (rates of infant mortality and years of average life expectation) are essentially measures of the adequacy of national health-care services, while the third (rates of adult literacy) assesses the adequacy of basic educational services. The PQLI makes no reference to other equally important spheres of social development activity, such as improvements in the changing status of women and children, protection of internationally guaranteed human rights, or participation in political decision-making. No valid index that attempts to measure changes in levels of national – and global – social development over time can exclude indicators of these critical areas of activity.

In constructing the PQLI, Morris did not attempt to produce a measure of "total welfare," but did indeed create an index that "promises to serve as a creative complement to the GNP" (Morris 1979, p. 93). It does, as Morris writes, "show how successful a country is in providing specific social qualities to its population and how it performs over time" (Morris 1979, p. 96). Even with serious limitations with respect to the range of social-development activities reflected in the index, the PQLI does contribute usefully to international development analysis and planning.

ADEQUACY OF NATIONAL SOCIAL PROVISION

Another approach to human welfare assessment is my own. Building on the social indicator work of the two UN Expert Groups (1954, 1961), McGranahan (1972), Drenowski (1970, 1974), Morris (1979), and others, I sought to construct a model of world *social-welfare* development that would reflect changes over time in the capacity of nations to provide for the basic social and material needs of their populations. Initially conceptualized as an "Index of National Social Vulnerability" (Estes & Morgan 1976), the assessment tool is now referred to as the "Index of Social Progress" (Estes 1984a & b, 1985).

In the present form the Index of Social Progress (ISP) consists of 44 welfare-relevant social indicators distributed among 11 subindexes: (a) education (4 items); (b) health (3 items); (c) status of women (5 items); (d) defense effort (1 item); (e) economics (4 items); (f) demography (5 items); (g) geography (3 items); (h) political stability (5 items); (i) political participation (6 items); (j) cultural diversity (3 items); and (k) welfare effort (5 items). The subindexes measure different dimensions of human welfare which, when considered together, constitute the major arenas for social-development activities nationally and internationally. The ISP itself measures what I refer to as "adequacy of social provision," i.e., the extent to which basic social needs are met within nations at discrete points in time.

The Index was field-tested in some 50 nations over a three-year period. Subsequently the ISP was applied to analysis of human-welfare changes that took place within 107 nations with 1969 populations larger than 1 million persons. Two time periods were selected to measure changes in adequacy of social provision that occurred within nations over time: 1969/70 and 1979/80.

The ISP was found to be a powerful tool for assessing changes in the capacity of nations to provide for the basic needs of their population during a critical decade of development worldwide. I reported separate analyses for each of the countries and, then, for groups of nations classified by economic development level and geographic location. The final chapters of the report group all nations together, thereby using the world as the unit of analysis.

Application of the ISP resulted in some findings that were not altogether expected at the outset of the research. Some of these findings have been controversial with respect to the relative positioning of particular countries – including that of the United States, which was found to rank 23 out of 107 nations on the study's Index of Net Social Progress (INSP) for 1980. Nevertheless, follow-up studies have continued to confirm the correctness of these earlier assessments (Estes 1986). In the main, the findings reported in *The social progress of nations* (Estes 1984b) are the result of the ISP's inclusion of a broader range of social phenomena that economically-oriented indexes exclude (e.g., the changing status of women and children, political forces, geographic and natural disasters, patterns of cultural diversity, and defense expenditures).

In my discussion of the implications of the study's findings I conclude that powerful economic and political forces exist which systematically serve either to promote or to retard patterns of social development worldwide. In the main, these internationally institutionalized forces work to the advantage of the world's already economically advanced nations and, in so doing, compound the developmental problems of less economically advanced nations. I also argue that these global social and economic forces function as determinants of global social stratification that push nations into specific "zones of social vulnerability." Vulnerability zones, I suggest, remain relatively stable over time and reflect the differing capacities of nations in achieving their developmental objectives. Table 3.1 reports changes in vulnerability-zonal positions for the 107 countries studied. (Note: Nations

Table 3.1 Distribution of nations into "zones of social vulnerability," using Index of Net Social Progress (INSP) scores, 1969–70 and 1979–80 (*N*=107).

	1980								
1970	Zone 1	Zone 2	Zone 3	Zone 4	Zone 5	Zone 6	Zone 7	Zone 8	Totals
Zone 1 (*X*=175+)*	Austria Denmark Ireland Netherlands Norway Sweden	FDR Germany							*N*=7 (6.5%)
Zone 2 (*X*=150–174)	Australia Finland New Zealand	Belgium Canada France Italy Japan Switzerland Bulgaria Hungary Poland Rumania	United Kingdom	Spain					*N*=15 (14.0%)
Zone 3 (*X*=125–149)		Czechoslovakia Costa Rica	Greece Portugal Yugoslavia Brazil Cuba Jamaica Panama Uruguay Venezuela	Albania USSR	Chile				*N*=14 (13.1%)

Table 3.1 *(cont.)*

					1980				
1970	Zone 1	Zone 2	Zone 3	Zone 4	Zone 5	Zone 6	Zone 7	Zone 8	Totals
Zone 4 (X=100–124)			United States Columbia Paraguay Trinidad/ Tobago	Argentina Dominican Republic South Korea Mexico Nicaragua Singapore Sri Lanka Thailand Tunisia Ecuador	South Africa Honduras Lebanon Philippines				N=18 (16.8%)
Zone 5 (X=75–99)				Turkey	Israel Algeria Bolivia Egypt El Salvador Guatemala Iran Malaysia Peru Haiti†	Burma Madagascar	Zimbabwe		N=14 (13.1%)
Zone 6 (X=50–74)				Syria	Libya Morocco	Cameroon Iraq Ivory Coast Kampuchea Kenya Liberia	Ghana Mauritania Zambia Somalia†		N=17 (15.9%)

						Senegal Sierra Leone Rwanda† Sudan†			
Zone 7 (X=25–49)					Indonesia	India Jordan Zaire Central African Republic† Guinea† Nepal† PDR Yemen†	Pakistan Togo Benin† Malawi† Mali† Niger† Upper Volta†	Burundi† Ethiopia† Tanzania† Uganda†	N=19 (17.8%)
Zone 8 [X=(–25)–24]						Vietnam	Nigeria	Chad†	N=3 (2.8%)
Totals	N=9 (8.4%)	N=13 (12.1%)	N=14 (13.1%)	N=15 (14.0%)	N=18 (16.8%)	N=20 (18.7%)	N=13 (12.1%)	N=5 (4.7%)	N=107 (100%)

* Indicates range of Index of Social Progress (ISP) scores used to define each range. Scores used in this distribution, however, are based on the Index of Net Social Progress (INSP) which excludes the Geographic Subindex scores from the aggregated ISP scores.

† Countries officially classified as "Least Developing Countries" (LDCs) by the United Nations.

Source: Richard J. Estes 1984. *The social progress of nations*. New York: Praeger.

located in Zone 1 are conceptualized as having the most favorable vulnerability position, while those located in Zone 8 are characterized as being the most "socially vulnerable," i.e., those in which the adequacy of social provision is at its lowest levels.)

As an approach to international human-welfare assessment, work on the ISP is not yet complete. Other indicators are being considered for inclusion in a revised version of the index (e.g., international trends in mental health, crime, and environmental pollution), and a new system of statistical weights is being considered. Nonetheless, the ISP, even in its present form, is more inclusive of a broader range of social phenomena related to "quality-of-life" than are other composite indexes of human welfare. The ISP also has the advantage of conceptual and methodological flexibility with respect to the assignment of nation- and region-specific norms. Further, the ISP does not assign greater saliency to a particular system of economic organization or political governance.

SUBJECTIVE APPROACHES TO HUMAN WELLBEING

In sharp contrast to the objective, highly statistical "basic needs" approaches taken by the world's major development assistance organizations, a number of investigators have sought to assess more subjective aspects of human wellbeing. Working in the early and mid-1970s, these researchers drew heavily from the psychological "needs-hierarchy" theories of Abraham Maslow (Maslow 1968). Their interest was in identifying the cross-cultural correlates of "felt satisfaction with life," "happiness," and "sense of personal security" (for instance, Kennedy *et al*. 1978, Andrews & Inglehart 1979). The investigators were concerned not so much with the extent to which the basic material needs of respondents were satisfied, but rather with the value they placed on their perceived satisfaction with life within whatever social context they found themselves.

Nearly all of the studies on the subjective aspects of human wellbeing, however, have been conducted with people living in a limited number of Western-oriented, democratic societies. Relatively few investigators have attempted to broaden their research focus to incorporate the majority of the world's population, who live in developing nations and cannot take for granted the satisfaction of basic social and material needs. The reason for this is obvious – the relative nontransferability (that is, nonequivalence) of basic psychological concepts from one culture to another. Interpretations of concepts such as "happiness" or "satisfaction with life," for example, differ significantly among people living within the same society and differ even more dramatically among people living in entirely different societies. The nontransferability of concepts from one culture to another has been a major stumbling block for all comparative investigators, irrespective of their philosophical inclination or methodological approach.

Relatively few large-scale comparative efforts on the subjective aspects of human wellbeing are currently in progress. For an especially comprehensive summary of the results of 245 studies of subjective "happiness" see the recent review completed by Ruut Veenhoven and his colleagues at the Erasmus University of Rotterdam (Veenhoven 1984).

Summary

In summary, then, a variety of empirical approaches have been attempted since the early 1940s to develop internationally valid indexes that both define and measure worldwide changes in the "quality-of-life" over time. Each approach to the problem has been unique but, overall, has drawn on the same rich tradition of empirical research in the social sciences. As a result, there is now a diversity of conceptual and methodological tools for studying changes in national and international human welfare. These analytical instruments have added measurably to the arsenal of resources that are needed to collect data for an increasingly expanding range of policy-relevant international social indicators. The research models developed thus far address numerous issues on the research agenda of quality-of-life researchers; but that agenda is far from being accomplished. Directions for future research in this internationally important area of social inquiry have been suggested throughout this chapter.

References

Andrews, F. & R. Inglehart 1979. The structure of subjective well-being in nine Western societies. *Social Indicators Research* **6**, 73–90.

Baster, N. & W. Scott 1969. *Level of living and economic growth: a comparative study of six countries*. Geneva: United Nations Research Institute for Social Development.

Bennett, M. K. 1951. International disparities in income levels. *American Economic Review* **41**, 632–49.

Davis, J. S. 1945. Standards and content of living. *American Economic Review* **35**, 1–15.

Drenowski, J. 1970. *Studies in the measurements of levels of living and welfare*. Report #70.3. Geneva: United Nations Research Institute for Social Development.

Drenowski, J. 1974. *On measuring and planning the quality of life*. The Hague: Mouton.

Estes, R. J. 1983. Education for international social welfare research. In *Education for international social welfare*, D. Sanders (ed.), Honolulu: University of Hawaii Press, 56–86.

Estes, R. J. 1984a. World social progress, 1969–1979. *Social Development Issues* **8**, 8–28.

Estes, R. J. 1984b. *The social progress of nations*. New York: Praeger.

Estes, R. J. 1985. Toward the year 2000: a social agenda for mankind. *Social Development Issues* **9**, 54–63.

Estes, R. J. 1988. *Trends in world social development*. New York: Praeger.

Estes, R. J. & J. S. Morgan 1976. World social welfare analysis: a theoretical model. *International Journal of Social Work* **19**, 29–41.

Kennedy, L. *et al.* 1978. Subjective evaluation of wellbeing: problems and prospects. *Social Indicators Research* **5**, 457–74.
Maslow, A. 1968. *Toward a psychology of being*, 2nd edn. New York: Van Nostrand.
McGranahan, D. *et al.* 1972. *Contents and measurement of socioeconomic development*. New York: Praeger.
Morris, M. D. 1979. *Measuring the conditions of the world's poor*. New York: Pergamon.
OECD 1973. *List of social concerns common to most OECD countries*. Paris: Organization for Economic Cooperation and Development.
OECD 1976. *The use of socio-economic indicators in development planning*. Paris: Organization for Economic Cooperation and Development.
OECD 1977a. *Measuring social well-being: a progress report on the development of social indicators*. Paris: Organization for Economic Cooperation and Development.
OECD 1977b. *1976 Progress report on phase II: plan for future activities*. Paris: Organization for Economic Cooperation and Development.
OECD 1981. *The welfare state in crisis: an account of the conference on social policies in the 1980s*. Paris: Organization for Economic Cooperation and Development.
OECD 1982a. *The challenge of unemployment: a report to labor ministers*. Paris: Organization for Economic Cooperation and Development.
OECD 1982b. *The OECD list of social indicators*. Paris: Organization for Economic Cooperation and Development.
OECD 1983. *Social expenditure statistics*. Paris: Organization for Economic Cooperation and Development.
OECD 1984. *National accounts of OECD countries*. Paris: Organization for Economic Cooperation and Development.
OECD 1985. *Social expenditures 1960–1990: problems of growth and control*. Paris: Organization for Economic Cooperation and Development.
Streeten, P. *et al.* 1981. *First things first: meeting basic human needs in developing countries*. New York: Oxford University Press for the World Bank.
UN 1954. *International definition and measurement of standards and levels of living*. Sales No. 1954.IV.5. New York: United Nations.
UN 1961. *An interim report on the international definition and measurement of levels of living*. Doc. No. E/CN.3/270/Rev. 1–E.CN.5/353. New York: United Nations.
UN 1964. *Handbook of household surveys: a practical guide for inquiries on level of living*. New York: United Nations.
UNESCO 1976. *The use of socioeconomic indicators in development planning*. Paris: United Nations Educational, Scientific and Cultural Organization.
UNESDP 1976. *Minimum levels of living*. Report No. 78–6650. Geneva: United Nations European Social Development Program.
Veenhoven, R. 1984. *Conditions of happiness*. Dordrecht & Boston: Reidel.
Verwayen, H. 1984. Social indicators: actual and potential uses. *Social Indicators Research* **14**, 1–27.

4 *Indexes of socioeconomic development*

ALFONSO GONZALEZ

The measurement of socioeconomic development

One of the most interesting and controversial problems in dealing with development is how to assess the level of development attained by individual countries; an associated problem is the assessment of comparative changes or progress among countries through time. The two most popular methods, among the many that have been proposed or used, that attempt to deal with these problems are the classifications based on gross national (or domestic) product (GNP or GDP) and the Physical Quality-of-Life Index (PQLI).

The GNP may be the most common method used to classify countries on a socioeconomic basis. This statistic has been utilized and popularized notably by the World Bank (*World Development Report*, various years). Although the classification of countries by the World Bank has undergone some modifications through time, the 1987 categorization was:

1 low-income economies (<$400)
2 middle-income economies ($400+)

   (a) lower middle-income ($400–1600) } developing countries
   (b) upper middle-income ($1600+) }

3 high-income oil exporters (not included in "developing countries")
4 industrial market economies
5 non-reporting non-member economies (most of the communist bloc)

The above categories are based, as they have been in earlier versions, essentially on income (actually per capita GNP) and the structure of the economy. The PQLI was devised by Morris David Morris and the Overseas Development Council (ODC) to measure the level-of-living on a scale of 0 to 100, and consists of a composite of three components (infant mortality, life expectancy at age 1, and literacy; Morris 1979).

Another measure is the Index of Social Progress (ISP) developed by Richard J. Estes (Estes 1984, Ch. 3 of this book). The ISP consists of 11 subindexes containing a total of 44 social, economic, physical, and political components. This index, therefore, is considerably more complex than the above two measures.

In my view, these measures are useful but contain inherent shortcomings: country rankings by these indexes do not conform to the world reality as I perceive them (discussed later in the Summary and Conclusions). The World Bank classification relies solely on per capita GNP and on the economic structure, i.e., whether oil-exporting or -importing, as well as whether the economy is market-oriented or centrally planned. This classification is used solely for the purpose of placing countries into groups based on a common structural characteristic. It relies on only one measure, GNP, and that particular measure has received considerable criticism (see below).

The PQLI uses three components, one of which (life expectancy at age 1) is not readily available for country or historical comparisons. That component with another (infant mortality) both measure the same factor, general health conditions. The third component (literacy) is probably not the best for indicating educational attainment. The net result is that when countries are ranked according to the PQLI (or according to the GNP alone), there are in my view too many disparities and the rankings do not reflect real or comparative levels of development.

The socioeconomic development index

In order to ascertain the level of socioeconomic development, or the level-of-living, I have felt the need to devise a comparative index (Gonzalez 1982). The components of the index were selected on the basis of their importance, their simplicity, the widespread availability of the data, and on the needs to avoid a special cultural bias, to measure performance (rather than inputs, effort, or priorities), to present a minimum of liability, to be capable of being used for historical comparisons (both relatively and absolutely), and to reflect the reality that the index is attempting to measure. The index that I have been using relies on four basic factors (income, diet, health, and education) that I feel are fundamental in development or in the level-of-living.[1]

Virtually all societies endeavor to improve or to increase all of these basic factors. There are proponents of a "no-growth" economy but these hardly reflect a significant or influential segment of any society. Some traditional societies may not emphasize income-expansion but such a view is not politically dominant anywhere, nor will policies based on it produce the resources to provide significant and rapid improvement in living conditions.

It can be assumed that improved diet and health are universal aspirations of all societies. The degree of education that is thought to be desirable varies between societies, but such desires may fall far short of what is required for an industrialized and technically advanced society. Since there is a universal desire for improvement in living conditions, I feel that the four factors cited best measure the degree of accomplishment and change through time ("development").[2]

INCOME

Income is based on per capita GNP. This statistic in the post-World War II period has become increasingly available. However, the use of the GNP as an indicator of income or development has often been criticized because of inaccuracy and incompleteness of coverage (notably in the underdeveloped world). Improvements have been made in correcting these deficiencies; nevertheless, objections continue. The criticisms are perhaps most often directed at the compilation of the GNP in the underdeveloped world. There are some omissions in compiling the GNP but these are generally of a minor order and would not greatly influence the overall per capita GNP for most countries. These local and subsistence producers are of low productivity and of small volume, and they are insignificant in the national economies, contributing little, if anything, to the national welfare. A compensating factor is that population may also be under-enumerated in this sector so that the per capita GNP would be little affected by these omissions. Further objections are raised with regard to the GNP as a measure of development. It is not an indicator of social conditions, but it is the best measure of income available. A larger per capita GNP does indicate that more funds are available for social welfare; but how that income is used, and how effectively, varies considerably among countries.

DIET

Diet is measured as a composite of kilocalories (kcal) and protein (grams) per capita daily consumption. These are clearly the best measures of nutrition for a country. Mean figures are used for countries, and the criticism can be levied that these mask a large proportion (especially in the underdeveloped world) of the population which consumes far less than the national average. This criticism is valid for this factor as well as for the other factors in this index – and just about everything in life. One assumption is fairly safe: all economic and social indicators of development must allow for a relatively small proportion of the population that is significantly *above* the national average and for a large mass that is *below* it.

Generally about 2,200 to 2,400 kcal and 60 grams constitute minimal levels of daily consumption. The minimum national daily requirements vary according to the age composition and the general body build of the population, and according to the climate. Underdeveloped countries might require somewhat less food per capita, but high-income sectors of those societies consume about as much as in the industrial countries. However, there are maximum limits to food consumption and these are probably being approached in the advanced world. There we can now observe various restrictions on caloric intake among an increasing share of the population. More may not be better when the national average exceeds about 3,000 kcal, but I prefer the *society* to set the limits (despite increasing evidence) rather than to establish my own maximum limit in the index. Certainly not all aspects of development can be considered as improvements.

There is, of course, a lower limit as well in food consumption. Undernutrition (with concomitant labor inefficiency) and eventually hunger and starvation result.

HEALTH

Health is represented by a composite of the infant mortality rate and life expectancy at birth. These indicators, like those for diet above, are now generally available and constitute the best available evidence of general health and medical conditions.

EDUCATION

Education is measured as a composite of literacy and the proportion of the population enrolled in higher (third-level) education. Literacy is used widely as an indicator of educational attainment, but has an important shortcoming in that its definition can vary significantly and constitutes only a first step in educational achievement for development. The proportion of the population in higher education may be questioned on the basis that many underdeveloped countries place greater emphasis on eradicating literacy than on rapidly expanding higher-educational facilities. This is of course true; but developed societies require technology and a higher order of learning. Furthermore, the index does not measure priorities but attainment and I have no doubt that technical training, research and development are major ingredients in the present and future "development" of societies.

The major country with the highest level of attainment in each of the above components (or sub-components) is given an index of 100, and all the others are given a figure based on the proportion of their individual performance as measured against the leader.[3]

THE INDEX AS A WHOLE

The overall index is the mean of the four basic factors, i.e., all four factors are given equal weight in determining the total overall index of socioeconomic development. It would be very difficult and subjective (depending on your own value judgements) to decide which of the components – income (expressed as per capita GNP), diet, health, education – is relatively more important in development than the others.

The major countries of the world were selected on the basis of their total population and GNP. In total 58 countries were chosen, slightly more than one-third of the world's sovereign countries (although one nonsovereign country, Puerto Rico, is included among the major countries).[4] These major countries account for slightly more than nine-tenths of the world's total population and GNP.

Components of the socioeconomic development index

The four components that comprise the overall total index each have equal weight. However, the components have different effects on the overall index. The component of income (per capita GNP) generally depresses the overall or total index. The major countries average 25 to 30 points lower in per capita GNP when compared with the overall index (Table 4.1). This is due to the great variation in countries' incomes, with great gaps possible; virtually all countries (except Saudi Arabia and Switzerland) have incomes much below the leader in comparison to other components. Furthermore, there appears to be little limit as to how much income can be attained, so the income range among countries may widen further.

In addition to significantly smaller incomes, the underdeveloped countries generally have a less egalitarian distribution of income. In the industrial countries (for which data are available), more than 6% of total income is accounted for by the poorest one-fifth of the population, while in the underdeveloped countries it is slightly more than 5% (based on data from *World development report*, 1987).[5] The richest 10% of the population in the industrial countries receive a quarter of the national income, whereas in the underdeveloped countries it is nearly two-fifths. In the underdeveloped world, the poorest group generally receives the smallest share of income in Latin America and the Middle East, while the rich receive a relatively smaller share of income in the Orient.

However, the uneven distribution of income is certainly not unique to the income component of the index. As indicated previously, virtually all characteristics (including the four components of this index) are very unevenly distributed not only worldwide but even within virtually all countries.

The diet component of the index is just the opposite of the income component: it inflates the overall index by approximately an average of 20 points. The reasons for this are fairly simple: food variability is not great because food consumption is necessarily limited (and there are some dietary restrictions in parts of the industrial world); also there is a lower threshold below which undernutrition, malnutrition, and ultimately starvation and death occur. This nutritional range is proportionately not very large.

In the last two components, health and education, overall levels generally approach most closely that of the overall index for socioeconomic development. Both consist of two elements each, with one in health (infant mortality) and one in education (literacy) having definite upper limits. However, in both of these components there are also significant variations

Table 4.1 Socioeconomic development index.

	Total*	per capita GNP†	Diet‡	Health§	Education¶
USA	90	86	96	78	100
Canada	82	73	89	83	83
Australia	74	66	83	79	70
New Zealand	71	45	96	73	71
USSR	57	16	90	54	66
UK	70	55	83	78	63
W. Germany	76	70	87	79	68
France	77	63	96	83	68
Italy	68	39	95	73	66
Netherlands	77	60	90	86	73
Belgium	73	56	95	75	68
Spain	65	29	87	80	65
Sweden	82	76	85	94	72
Switzerland	85	100	87	90	62
E. Germany	69	31	97	76	71
Poland	60	23	92	62	63
Czechoslovakia	63	34	91	66	62
Romania	54	12	90	57	57
Yugoslavia	57	16	95	55	60
Mexico	46	14	73	49	49
Cuba	53	10	74	66	62
Puerto Rico	57	24		67	80
Venezuela	50	25	65	53	57
Colombia	41	9	59	47	50
Peru	37	6	55	41	45
Chile	49	11	69	59	57
Argentina	56	12	95	54	62
Brazil	39	12	61	45	39
Morocco	33	5	66	41	21
Algeria	36	15	65	42	22
Egypt	39	4	78	41	33
Sudan	27	2	60	34	12
Iraq	40	14	71	43	32
Saudi Arabia	51	74	76	39	15
Israel	64	33	85	70	67
Turkey	42	8	78	44	40
Iran	38	13	72	42	26
India	28	2	49	37	24
Pakistan	27	2	55	35	15
Bangladesh	23	1	43	33	15
Burma	35	1	59	39	41
Thailand	41	5	52	47	59
Vietnam	34	1	49	46	41
Malaysia	42	11	63	54	41

Table 4.1 (*cont*).

	Total*	per capita GNP†	Diet‡	Health§	Education¶
Indonesia	32	3	54	39	33
Philippines	44	5	56	48	67
China	34	2	60	50	26
Taiwan	53	13		82	64
S. Korea	51	12	76	54	64
Japan	77	62	79	100	69
Nigeria	28	5	57	35	15
Ghana	26	2	47	37	18
Zaire	24	1	44	35	18
Ethiopia	25	1	61	30	6
Kenya	27	2	52	38	17
Tanzania	31	1	56	36	32
Uganda	25	1	45	37	17
South Africa	40	15	72	38	35
Advanced	71	50	90	76	69
Underdeveloped	38	9	62	45	36

* The average of the four factors (per capita GNP, diet, health, and education).

† Based on the *World Bank Atlas* (1985 and earlier editions).

‡ Based on the daily per capita consumption of kilocalories and protein, from the *FAO Production Yearbook* (1984 and earlier editions).

§ Based on the infant mortality rate and the life expectancy at birth, from the *World population data sheet* (1985).

¶ Based on literacy and the proportion of the population in higher (third-level) education, from the UNESCO *statistical yearbook* (1983 and earlier editions).

among individual countries. Data for the other element in the health component (life expectancy at birth), like all the components of the index, are readily available for almost all countries, and this element closely rivals the infant mortality rate as the best indicator of health conditions.

The literacy component of education has an upper limit (100% literacy) and very little discrimination is possible among advanced countries. Furthermore, there is also the question of what constitutes "literacy" and definitions vary among countries, but this component is important if one considers the emphasis that many countries place on campaigns to eliminate illiteracy. The other component of education (proportion of the population in third-level or higher education) is a good measure of the degree of training occurring at a high level. No doubt one can argue about the differences among countries in the "quality" of education being dispensed. It is also true that students studying abroad are not counted and foreign students within the country are included.

There is considerable variation in almost all countries when one compares the individual components of the index. In most countries at least two components (frequently three, especially in the underdeveloped world) will vary at least 10 points from the overall index. Saudi Arabia varies greatly in all four components whereas Canada and Australia vary little in any of them. West Germany, the USA, and Sweden are the only countries that vary significantly in only one component. Therefore, it would appear that some countries rank relatively much higher or lower in some aspects of development than in others.

PER CAPITA GNP

Considering the general depression of this component with regard to the overall index and the overall level of socioeconomic development attained by the individual countries, per capita GNP is generally higher than expected in the countries of the First World – i.e., Western Europe, Anglo-America (USA and Canada), and Japan, and Oceania – and in some of the OPEC countries of the Middle East. This is especially true of Saudi Arabia, Switzerland, the USA, Sweden, West Germany, Australia, and Canada. However, incomes (or per capita GNP) are significantly *lower* in the countries of Eastern and Southern Europe, the USSR, and especially Argentina and Cuba, which rank much below what would be expected.

DIET

In some countries dietary levels are higher than their levels of development would indicate, notably in portions of the industrialized Communist bloc and the Middle East. The individual countries that rank relatively very high are Argentina, Egypt, Yugoslavia, Ethiopia (perhaps), Romania, and Turkey. Dietary levels are somewhat low in several countries of the First World (due essentially to the upper dietary restriction, especially in Japan, Switzerland, and Sweden) and scattered countries of the underdeveloped world.

HEALTH

Variations are generally less extreme, but Saudi Arabia, the USA, and the USSR do rank below expectations. The relatively high-ranking countries are Taiwan (Republic of China), Japan, China (People's Republic of China), Spain, and Cuba.

EDUCATION

In this component, as with health, the index generally approaches the overall index of development. The relatively very high-ranking countries are Puerto Rico, the Philippines, and Thailand, while the relatively very low-ranking countries are those of the Middle East, subSaharan Africa, and Western Europe (notably Saudi Arabia, Switzerland, Ethiopia, and Sudan).

In general, on a regional basis, Western Europe and Anglo-America rank relatively rather high (considering the other components of socioeconomic development) in income; whereas Eastern Europe, the USSR, and Latin America have relatively depressed incomes for their overall level of development. However, in diet it is Western Europe and Anglo-America that rank little above their level of development (although they are the best-fed regions), while Eastern Europe, the USSR, and the Middle East have a dietary level significantly higher than their overall level of development. In health, regional variations are not very great, and the same is true to a lesser degree of education, although the index for Latin America is somewhat advanced and the Middle East held back by the factor of education.

Levels of socioeconomic development

As one would expect there is a very gradual change in the level of development among the countries of the world. Using only the 58 major countries in this study, differentiation between levels of socioeconomic development is sometimes difficult and arbitrary or subjective. Furthermore, changes in rankings are also gradually occurring. Nevertheless, perhaps six levels of development can be identified among the major world countries on the basis of this socioeconomic development index (Table 4.1) and rankings during the 1970s.

(1) highly developed (indexes of 90–73): the countries of Anglo-America, most of Northwestern Europe, Japan, and Australia;
(2) advanced development (71–60): the UK, Southern Europe, the northern countries of Eastern Europe, the USSR (perhaps marginally), New Zealand, and Israel;
(3) transitional (57–46): southern Eastern Europe (and perhaps the USSR), two-thirds of the Latin American major countries (Puerto Rico, Cuba, Venezuela, Chile, Mexico), South Korea, and Saudi Arabia;
(4) less developed (44–37): the remainder of Latin America, the most populous countries of the Middle East (Turkey, Egypt, Iran) plus Iraq, and the Philippines, Malaysia, and Thailand in the Orient, and South Africa;
(5) underdeveloped: (36–32): the Maghreb of Africa (Algeria and Morocco), and Burma, China, Vietnam, and Indonesia;
(6) very poor/least developed (31–23): Sudan, the Indian sub-continent, and all of subSaharan Africa except South Africa.

In overall development on a world regional basis, Anglo-America is clearly the most advanced followed by Western Europe and Australia–New Zealand. Significantly behind these are Eastern Europe and the USSR, followed by another comparably significant break and then Latin America. Another major break separates Latin America from the Middle East and the Orient. Significantly further back is subSaharan Africa.

Table 4.2 Socioeconomic development index for major countries.

Per capita GNP ($)*	1983	PQLI†		ISP‡	1979	SEDI§	1980s
Switzerland	16,390	Sweden	97	Netherlands	190	USA	90
USA	14,090	Netherlands	96	Sweden	189	Switzerland	85
Sweden	12,400	Japan	96	New Zealand	186	Canada	82
Saudi Arabia	12,180	Canada	95	Australia	184	Sweden	82
Canada	12,000	Switzerland	95	Belgium	178	France	77
W. Germany	11,420	USA	94	W. Germany	174	Netherlands	77
Australia	10,780	New Zealand	94	Canada	170	Japan	77
France	10,390	UK	94	Switzerland	170	W. Germany	76
Japan	10,100	France	94	Poland	168	Australia	74
Netherlands	9,910	Australia	93	France	165	Belgium	73
Belgium	9,160	W. Germany	93	Czechoslovakia	163	New Zealand	71
UK	9,040	Belgium	93	Romania	163	UK	70
New Zealand	7,410	E. Germany	93	Italy	158	E. Germany	69
Italy	6,350	Czechoslovakia	93	Japan	149	Italy	68
Czechoslovakia	5,585	Italy	92	UK	145	Spain	65
Israel	5,360	USSR	91	Cuba	141	Israel	64
E. Germany	5,100	Spain	91	Yugoslavia	137	Czechoslovakia	63
Spain	4,800	Poland	91	Venezuela	137	Poland	60
Venezuela	4,100	Romania	90	Brazil	137	Puerto Rico	57
Puerto Rico	3,890	Puerto Rico	90	Colombia	130	USSR	57
Poland	3,710	Israel	89	Spain	129	Yugoslavia	57
USSR	2,605	Taiwan	86	Argentina	124	Argentina	56
Yugoslavia	2,570	Argentina	85	Mexico	121	Romania	54
South Africa	2,450	Yugoslavia	84	USA	116	Taiwan	53
Algeria	2,400	Cuba	84	USSR	113	Cuba	53
Iraq	2,300	S. Korea	82	Turkey	112	S. Korea	51
Mexico	2,240	Venezuela	79	S. Korea	107	Saudi Arabia	51
Taiwan	2,143	Chile	77	Thailand	99	Venezuela	50
Iran	2,123	Mexico	73	South Africa	98	Chile	49
Argentina	2,030	Colombia	71	Algeria	96	Mexico	46
S.Korea	2,010	Philippines	71	Israel	92	Philippines	44
Romania	1,913	Brazil	68	Malaysia	92	Malaysia	42
Brazil	1,890	Thailand	68	Philippines	91	Turkey	42
Chile	1,870	Malaysia	66	Chile	90	Colombia	41
Malaysia	1,870	Peru	62	Egypt	81	Thailand	41
Cuba	1,592	Turkey	55	Peru	76	South Africa	40
Colombia	1,410	Vietnam	54	Morocco	73	Iraq	40
Turkey	1,230	South Africa	53	Indonesia	71	Brazil	39
Peru	1,040	Burma	51	Burma	70	Egypt	39
Thailand	810	Indonesia	48	Iran	69	Iran	38
Philippines	760	Iraq	45	Iraq	66	Peru	37
Nigeria	760	Egypt	43	Sudan	60	Algeria	36
Morocco	750	Iran	43	Kenya	56	Burma	35
Egypt	700	India	43	Vietnam	53	China	34
Indonesia	560	China	43	India	53	Vietnam	34
Sudan	400	Morocco	41	Zaire	52	Morocco	33
Pakistan	390	Algeria	41	Ghana	39	Indonesia	32
Kenya	340	Uganda	40	Nigeria	33	Tanzania	31
Ghana	320	Kenya	39	Pakistan	31	India	28
China	290	Pakistan	38	Tanzania	29	Nigeria	28
India	260	Sudan	36	Uganda	21	Kenya	27
Tanzania	240	Bangladesh	35	Ethiopia	−12	Sudan	27
Uganda	220	Ghana	35	E. Germany¶		Pakistan	27
Burma	180	Zaire	32	Puerto Rico¶		Ghana	26
Vietnam	160	Tanzania	31	Taiwan¶		Uganda	25
Zaire	160	Saudi Arabia	29	China¶		Ethiopia	25
Ethiopia	140	Nigeria	25	Bangladesh¶		Zaire	24
Bangladesh	130	Ethiopia	20	Saudi Arabia¶		Bangladesh	23

* Based on *World Bank atlas* (1985 and earlier editions).
† Physical Quality-of-Life Index. Based on Morris 1979.
‡ Index of Social Progress. Based on Estes 1984.
§ Socioeconomic Development Index. Based on Table 4.1.
¶ Not evaluated by this index.

Summary and conclusions

Indexes or measures of socioeconomic development (quality of life or human welfare) may vary from simple and statistically unsophisticated to comprehensive, multidimensional and statistically and methodologically rigorous. One would presume that with a greater number of indicators and increasingly sophisticated statistical analysis the index would be more valid and acceptable. The real test of these indexes, however, is how truly they reflect real world conditions and, in this, subjectivity is a consideration.

The four indexes discussed in the early sections of this chapter can be compared with regard to the world's major countries and regions (Table 4.2). Although the indexes were compiled on data from the mid-1970s to the early 1980s the differences when compared to the mid-1980s should not be very great.

When the four indexes are compared for the major countries, nearly two-fifths of the countries deviate significantly from the norm (of the four indexes) in at least one index. Three countries (Venezuela, South Africa, and Saudi Arabia) especially appear troublesome to categorize. Venezuela and South Africa are ranked high by both the GNP and Index of Social Progress (ISP), while the Physical Quality-of-Life Index (PQLI) and the Socioeconomic Development Index (SEDI) rank them considerably lower. Saudi Arabia is ranked very high by GNP and very low by the PQLI, while the SEDI ranks it just about halfway between both. The ISP does not evaluate Saudi Arabia as well as five other "major" countries. Romania also presents somewhat of a problem as the ISP ranks it very high (well above Japan, the UK, and the USA) whereas the GNP ranks it very low. The PQLI and the SEDI rank it midway between these two indexes.

The use of the per capita GNP alone in ranking the major countries does bring out one major point. Since the mid-1970s when the price of petroleum began its rapid rise, the petroleum-rich countries (notably of the Middle East) have markedly increased their per capita GNP. As a consequence, some major petroleum producers, especially Saudi Arabia, Algeria, Iraq, and Iran, rank much higher in this index than in any of the others. Some countries, notably the Netherlands, Burma, and the Communist bloc (especially Romania and Vietnam), rank much lower than expected when using only the per capita GNP. Measurement of the GNP in Communist (non-market or centrally planned) economies do present obstacles to measurement and the World Bank has apparently been grappling with this problem.

For the other indexes (in addition to the countries mentioned above) the PQLI ranks the UK and the USSR somewhat higher than the other indexes, while Nigeria comes out lower. The ISP ranks a number of countries somewhat higher than the other indexes, notably Belgium, Poland, Cuba, Colombia, Brazil, Turkey, and Zaire; while downgrading others, notably the USA (one of the great differentials), Switzerland, Israel, and Japan.

The use of the socioeconomic development index (SEDI) is in my view sufficiently sophisticated, without being unduly complex and cumbersome, to allow adequate discrimination between different levels of development. Furthermore, it comes closer in my view to satisfying the ranking of countries with my own observations of the real world.

Different countries have had greater success in one field of socioeconomic development than in others and many countries exhibit significant variation from one component to another. There are many variations in the degree of "balanced development" among countries.

There are also significant variations in development between the world's major regions and within each region itself: I hope that this index makes those differences clearer.

Notes

1 If the purpose is to measure the quality of life, I add a fifth factor, political freedom (as measured by Freedom House in New York).

2 Most of these components, if not all, are occasionally called into question because they are allegedly inaccurate, incomplete, or misleading. However in my estimation these are probably the best available and, although not precisely accurate, they can be used for comparative purposes. GNP data for the communist bloc is being reviewed by the World Bank and provides limited data for these and some other countries. I have presented the latest estimates available for those countries and others for which the most recent data are not available.

3 Since there are two components in the measures of diet, health, and education, it is possible that the same country may not be the leader in both components of the same factor. This would result in no country with a ranking of 100 for that factor. This is what occurs for diet and possibly education.

4 Comparable data for both Puerto Rico and Taiwan, especially with regard to diet, are not available. Therefore in computing the overall index, both countries include only three of the factors instead of four. However, in diet both countries would rank fairly high, by the standards of the underdeveloped world.

5 Data are available for a total of 52 countries (35 of which I have classified as "major" in this study). All countries have been utilized in this brief discussion of income distribution.

References

Estes, R. J. 1984. *The social progress of nations*. New York: Praeger Publishers.

FAO production yearbook. 1984 (and earlier editions). Rome: Food & Agriculture Organization.

Gonzalez, A. 1982. A measure of the level of living or the quality of life. Unpublished paper. Association of American Geographers, San Antonio, Texas.

Morris, M. D. 1979. *Measuring the condition of the world's poor*. Washington: Pergamon & Overseas Development Council.

Sivard, R. L. 1985 (and earlier editions). *World military and social expenditures*. Leesburg, Va.: World Priorities.

UNESCO *statistical yearbook.* 1983 (and earlier editions). Paris: United Nations Educational, Scientific & Cultural Organization.
World Bank atlas 1985 (and earlier editions). Washington: World Bank.
World development report 1985 (and earlier editions). Washington: World Bank.
World population data sheet 1985 (and earlier editions). Washington: Population Reference Bureau, Inc.
World population estimates 1981 (and earlier editions). Washington: The Environmental Fund.

PART II

Selected Themes and Patterns

God has pity on kindergarten children.
He has less pity on school children.
And on grownups he has no pity at all,
He leaves them alone,
and sometimes they must crawl on all fours
in the burning sand
to reach the first-aid station
covered with blood.

Yehuda Amachai.
"God has pity on kindergarten children",
In "Israel's master poet", Robert Alter,
New York Times Magazine, June 9, 1986.

This part provides analyses of some of the diverse problems and themes that characterize the process and practice of development. The topics and themes selected are of necessity limited, since it is not possible to provide a truly comprehensive coverage of such a complex question as development and the Third World.

John Cole provides a description and discussion of the distribution of the world's natural resources (bioclimatic, fuels/energy, and non-fuel minerals). His analysis leads to the conclusions that elimination of political constraints would have little overall impact and that there has been little change in recent decades between the rich and poor countries.

Robert Bednarz and **John Giardino** begin with a discussion of the methods of evaluating the quality of life, and the problems associated with indexes, including lack of agreement as to variables and the cross-cultural differences. They provide another measure of the quality of life (including the resource base), which, like Richard Estes', is somewhat more inclusive and complex than others. They find virtually no relationship between the quality of life and the national resource base and they suggest that natural resources are more important in the earlier stages of development.

Gerald Ingalls and **Walter Martin** discuss the various definitions and classification of the "newly industrialized countries" (NICs). They establish five criteria and determine that there are three countries, and seven others possibly, that are NICs. They then provide more detailed country analyses.

Laurence Wolf, in "The poorest of us all," provides a discussion of the Third World from the core–periphery perspective. He classifies countries into categories on the basis of per capita GNP and the Physical Quality of Life Index (PQLI). The author finds little recent relative improvement in per capita GNP for Third World countries; some improvement in the physical quality of life has occurred, but generally it is not related either to GNP or to earlier PQLI values. He also relates PQLI values to population growth rates and finds some correlation between improved quality of life in the Third World and slower population growth, although there are some significant exceptions. He concludes that improvements in the Third World are inadequate to close the gap with the core (the industrial or developed countries) and that prospects for doing so are not good.

Jerome Fellmann examines health conditions in the world and finds that comparable per capita GNP does not assure comparable health and that low income does not necessarily imply ill health. Income distribution within a country is more significant. He associates the stages of the demographic transition with a parallel epidemiologic transition and points out that Third World patterns do not duplicate the Western model. He describes two useful indexes and evaluates Third World (and global) health conditions and diseases. He feels that the surest guarantee for health improvement is literacy (especially among females) and that there is gradual progress towards advanced health standards.

Alice Andrews examines the status of women in the Third World. She evaluates this in terms of the law, prestige, and overall conditions and proposes that it should also include independence and access to resources. Although there are difficulties in obtaining comparable data, she examines a variety of indicators, essentially education and economic and demographic elements. She provides regional summaries for the developing areas. Women's status is generally low, but variable, with the best conditions in Latin America and perhaps the most disadvantaged in subSaharan Africa. She feels that the greatest hope for improvement is in education, perhaps even more so than in increased income.

Laurence Wolf and **Thomas Anderson** analyze the concept of freedom. Laurence Wolf indicates the conditions of the Third World that restrict the development of political freedom and presents a varied set of problems. He indicates that in the Third World the evolution of civil liberties and democracy has not parallelled economic growth. He briefly outlines the great variety of political regimes of the Third World and points out that none has a long uninterrupted democratic tradition (such traditions are confined to the First World). In his view only four countries of the Third World have some democratic tradition (temporary/transitory democracies are not included), and although traditional bourgeois democracy is not the only possible model, there is no better alternative. Capitalism, however, is not equated with democracy and the export of political freedom has not been successful.

Thomas Anderson discusses the controversy over human rights and the world covenants concerning these, and notes the differences between the interpretations of these covenants by the First World compared with those by the Second and by many countries of the Third World. He feels that these different interpretations are not mutually exclusive, but he clearly supports the Western view of political and individual choices. He outlines the principles of the Democratic Revolution and points out that these ideas are not restricted to West European culture. He ranks the countries of the world according to the degree of personal liberty, and analyzes conditions within the six categories. He finds that Africa is the least free; that there are few repressive governments in the Western Hemisphere; and that functioning democracies are not necessarily associated with highly literate populations and advanced economies.

Alfonso Gonzalez discusses trends in socioeconomic development among the major world countries and regions during the decades of the 1960s and 1970s. The comparisons are both in absolute and relative change (relying on his index). Industrialized countries slowed their growth during the recent world recession but have improved at least as much as the Third World. In the latter, the most severe problems are in subSaharan Africa, with the Middle East demonstrating the greatest recent improvement, and Latin America being at the highest level. Many countries are more successful in one (or more) fields of development than in others, and closing the gap with the advanced world will probably be slow at best.

A. Gonzalez

5 *The global distribution of natural resources*

JOHN P. COLE

Introduction

For the purpose of this chapter the term *natural resources* denotes the set of all elements that exist in the world regardless of the presence of humans but that are currently or potentially of use to humans. Apart from flows of air and water, natural resources are largely fixed in location. Goods produced from them can be transported elsewhere and people can move to them or from them. For simplicity natural resources may be grouped as follows (only the first two will be considered here):

Bioclimatic resources: the land surfaces and water bodies from which plant and animal products are derived. Although water has a variety of uses, in view of its great importance to agriculture it will be included here.

Fossil-fuel and non-fuel mineral resources, subdivided according principally to function into fossil fuels and nonfuel minerals.

Others which do not readily fall into the first two categories, such as solar energy and its effects (e.g., falling water as a source of electric power), and also geothermal power.

Bioclimatic resources are commonly regarded as non-exhaustible, while mineral resources are exhaustible (or nonrenewable) – an over-simplified view, if a convenient one. In reality productive land, whether arable, pasture or forest, is nonrenewable once it is lost through soil erosion or through the construction of buildings, roads and other works of man, as Brown (1978) vividly illustrates in *The worldwide loss of cropland*. In contrast, some new mineral deposits are forming, but generally at rates much lower than the current rates of their extraction. Another contrast between bioclimatic and mineral resources is the fact that many parts of the world have only been explored superficially for minerals whereas the quality and potential productiveness of most of the earth's land surface is broadly known.

The products from the various natural resources are to some extent in competition. Fossil fuels are also the source of valuable raw materials (synthetic materials such as plastics and rubber substitute), while bioclimatic products are a major source of fuel and power (e.g., fuelwood, dung, and

Table 5.1 Distribution of major types of natural resources in relation to population, for 20 countrie

Country	Population* (1982) ('000)	Bioclimatic (early 1980s) arable† (ha/100 inhabitants)	forest‡ (ha/100 inhabitants)	Fossil fuels (early 1980s) coal and lignite§ (tonne/ inhabitant)	oil¶ (tonne/ inhabitant)	natural gas** ('000 m³/ inhabitant)	Non-fuel minerals†† (per inhabit world avera =100)
1 China	1,021	10	13	98	3	1	10
2 India	721	24	10	10	1	1	6
3 USSR	270	86	341	648	31	152	258
4 USA	232	82	122	823	19	24	125
5 Indonesia	153	13	80	0	8	7	29
6 Brazil	128	58	445	55	2	0	107
7 Japan	118	4	21	8	0	0	12
8 Bangladesh	93	10	2	0	0	2	0
9 Pakistan	92	22	3	0	0	4	0
10 Nigeria	82	37	17	0	28	12	11
11 Mexico	74	32	64	14	92	30	113
12 German FR	62	13	11	774	0	3	58
13 Italy	57	21	11	0	0	0	8
14 Vietnam	56	11	18	0	0	0	0
15 UK	56	13	4	89	32	13	17
16 France	54	35	28	0	0	0	25
17 Philippines	52	23	23	0	0	0	75
18 Thailand	49	39	33	0	0	0	0
19 Turkey	48	52	42	0	0	0	0
20 Egypt	44	6	0	0	9	4	0
20 countries	3,453	29	66	155	8	16	–
World	4,593	32	89	157	21	21	100

Sources: *, †, ‡ *Food and Agriculture Organization production yearbook 1983*, Tables 1, 3.
§ United Nations *1982 Energy statistics yearbook*, Table 43.
¶, ** *BP statistical review of world energy*, various tables.
†† *Mineral facts and problems 1980*, various tables.

alcohol). The sources from the category "Other" would provide the bulk of the "inanimate" energy used in the world if and when this resource is widely used. A further link in the "system" is the dependence on metallic nonfuel minerals to provide the means of production to carry out many productive processes, and of nonmetallic nonfuel minerals to provide fertilizers to increase yields in agriculture.

The main purpose of this chapter is to describe the distribution of natural resources in the world in relation to population. Attention will be drawn to the enormous inequalities that exist in this respect, particularly among Third World countries but also with reference to the world as a whole. The abundance or lack of particular natural resources in developed countries greatly influences their attitude to Third World countries and the structure of international trade.

The data in Table 5.1 show for the 20 countries of the world with the largest populations the availability of major types or groups of natural resources per inhabitant in the early 1980s. Population is for 1982 because at the time of writing the latest available data were for around that year. A zero in the table may signify either nothing or a negligible amount.

Bioclimatic resources

Official figures, many of which are only intelligent estimates, are given in the *Food and Agriculture Organization Production Yearbooks*, for four major types of land-use in each country of the world. The four types of land-use (or lack of use) are: "arable," "permanent pasture," "forest and woodland," and "other." In the arable type a subsection of permanent crops (such as coffee and vines) is distinguished, and fallow is also included.

The productivity of arable land varies greatly from place to place according to soil and climatic conditions, capital inputs and techniques. In some parts of the world its usefulness can be greatly enhanced by irrigation. Much arable land is double-cropped, especially in warmer parts of the world. For example, from roughly 100 million hectares of cropland in China about 150 million hectares of crops are harvested each year. Elsewhere, especially in areas of shifting cultivation, much "arable" land is out of use for long periods. For comparison, the availability of bioclimatic resources in relation to population is shown in Table 5.1 for both developed and developing countries.

The data in the second column of Table 5.1 show the area of arable land in relation to population. In the world as a whole there were on average about 30 hectares per 100 inhabitants, but even among the 20 countries of the world with the largest populations the contrast between Japan, with only about four hectares, and the USSR with 86 hectares is striking. Australia has about 315. Marked contrasts among Third World countries are shown in more detail in Table 5.2.

For comparison with arable land, the number of hectares of forest per 100 inhabitants is shown in the third column of Table 5.1. The contrast between Brazil and Egypt might be expected but the lack of forest in South Asia and China is perhaps surprising. The forests of the developed countries are being depleted gradually but it is in many developing countries, especially those in the drier parts of Africa and in much of Asia, that the problem is most acute as in those countries wood is vital as a fuel.

Of particular importance to development studies is the fact that the area of arable land in the world is increasing less rapidly than population. Reference to this fact often provokes an impatient or even hostile reaction. At present only about 11% of the world's land area is cultivated. There must be plenty of room for expansion. One cannot assume automatically, however, that

Table 5.2 Changes in the availability of arable land per inhabitant in selected larger developing countries.

	Population ('000,000)		Arable land and permanent crops ('000,000 ha)		(ha/100 inhabitants)				
	1951	1982	1951	1982	1951	1963	1975	1982	2000
Mexico	27	74	19	24	69	62	39	32	29
Brazil	54	128	20	75	37	40	56	58	48
Colombia	12	27	3	6	22	30	23	21	17
Nigeria	34	82	22	30	66	48	46	37	23
Ethiopia	17	33	11	14	66	55	48	43	27
Morocco	9	22	8	8	86	56	45	39	24
Egypt	21	44	3	3	12	9	8	6	4
Iran	17	41	17	14	101	92	50	34	22
Pakistan	34	92	16	20	47	36	26	22	15
India	363	712	138	170	38	35	27	24	18
Bangladesh	42	93	8	9	19	16	12	10	6
Indonesia	77	153	11	20	14	17	14	13	11
Philippines	21	52	5	12	22	23	23	23	19
China	563	1,021	95	101	17	14	11	10	9

Main source: Food and Agriculture Organization Production Yearbook (various years).

land now in the "natural pasture" and "forest" categories of land-use are simply there to become cropland one day; and certainly very little of the fourth category, "other" (which is largely waste), could be used. It is often pointed out also that yields of the same crop are much lower in some countries than in others and, with improvements (especially through the greater use of chemical fertilizers), production could in theory be greatly increased in many areas. Even so, it is worth noting trends in a selection of developing countries to see what is actually happening with regard to the availability of land that is currently arable.

In Table 5.2 estimates for 14 selected developing countries, mostly large in population, show that the ratio of area of arable land to population decreased sharply between 1951 and 1982 in ten countries, changed little in three (Colombia, Indonesia, and the Philippines), and increased only in one (Brazil). These last four all contain extensive areas of forest.

In general the increase in arable area has been slower than the growth of population. A comparison between the estimated population for the year 2000 and arable area, after assuming some increase to that date, shows that in most countries the situation is likely to deteriorate considerably in the next two decades or so. In some countries, exemplified in Table 5.2 by Egypt, there is just no slack to take up. When the Communist Party gained power in China in 1949 there were about 18 hectares of arable per 100 people. By about 2030, if the increase of population in China recorded in the 1950s and

Table 5.3 Consumption of commercial sources of energy in selected developed and developing countries.

Country	Energy consumption in 1982 (kg coal-equivalent/ inhabitant)	Country	Energy consumption in 1982 (kg coal-equivalent/ inhabitant)
USA	9,430	Egypt	620
USSR	5,770	China	580
German FR	5,510	Thailand	360
UK	4,540	Philippines	330
France	4,000	Indonesia	230
Japan	3,500	Pakistan	210
Italy	2,890	Nigeria	200
Mexico	1,760	India	200
Turkey	790	Vietnam	130
Brazil	700	Bangladesh	50

Source: *1982 Energy statistics yearbook* (United Nations).

1960s were to continue, there would only be about six hectares per 100 people – one reason used by Liu Zheng (1981) to justify the need to enforce a drastic family planning policy.

In many developed countries there has been little change in the arable area in recent decades; and in some there has been a loss due to urbanization and to the policy, especially in the European Economic Community, of withdrawing marginal land from cultivation. In Japan there has been roughly a 20% drop in the cultivated area in the last three decades. Developed countries are however better able than developing countries to increase yields, and to import food and other agricultural products, either from other developed countries or from developing countries, while population growth is generally slow. Many developing countries are now reluctant net importers of food though many maintain a net export of beverages and plant raw materials.

Fossil fuels and other sources of energy

It is widely appreciated that the level of consumption of inanimate sources of energy varies greatly among the countries of the world (see Table 5.3). The data in Table 5.3 show contrasts that are perhaps greater than imagined and in particular reflect enormous differences between the levels of use of machinery of various kinds in different parts of the world.

For many decades now fossil fuels have provided a very large proportion of the total fuel and power used in the world. In the early 1980s some 9,000 million tonnes of coal-equivalent of commercial energy were consumed each

Table 5.4 Reserves of coal and lignite, by country.

Country	Coal: recoverable* (10^9 tonne)	Coal: in place† (including recoverable) (10^9 tonne)	Coal: additional‡ (10^9 tonne)	Country	Lignite recoverable (10^9 tonne)
1 USA	125	224	472	1 USA	132
2 USSR	109	136	1,710	2 USSR	132
3 China	99	200	1,326	3 East Europe§	65
4 South Africa	52	112	17	4 Australia	38
5 German FR	30	44	186	5 German FR	35
6 Australia	27	49	507	6 Brazil	13
7 Poland	27	60	84	7 Canada	4
8 UK	5	n.a.	185		
9 India	n.a.	26	86		
10 Mongolia	n.a.	12	n.a.		
11 Botswana	4	7	100		
12 Canada	2	n.a.	26		
World	515	n.a.	n.a.	World	430

Source: Mainly *1982 Energy statistics yearbook* (United Nations), Table 43.
n.a. Not available.
* "Recoverable": the mineral can be extracted commercially.
† "In place": sources known to exist but from which the mineral is not currently recoverable.
‡ "Additional": possible reserves not yet explored fully.
§ Bulgaria, Czechoslovakia, German DR, Hungary, Poland, Yugoslavia.

year, around 1,800–1,900 kilograms (nearly two tonnes) per inhabitant. In 1982, 30% was accounted for by coal and lignite, 45% by oil, and 21% by natural gas. The rest came mostly from hydroelectric and nuclear-generated electricity.

In developing countries, with much lower levels of energy consumption per inhabitant than those in most developed countries, the use of wood for fuel (defined however as "non-commercial") is very widespread, and in 1981 amounted to some 560 million tonnes of coal-equivalent. This energy-equivalent was obtained from the cutting of some 1,600 million cubic meters of wood, compared with the cutting of 3,000 million cubic meters of roundwood for construction and other non-fuel purposes. The world's forests and savanna lands are thus threatened as a source of fuel, raw materials and space for extending cultivation. The use of working animals in developing countries is another considerable source of non-commercial energy. Which countries, then, have the fossil fuels?

The countries with the largest reserves of coal and lignite in the world are listed in Table 5.4. There is no standard method of assessing coal and lignite reserves so estimates for some classes and countries are not available. The countries of the world are listed roughly according to the size of their reserves. The astronomical size even of the recoverable reserves of coal can be appreciated from the fact that at 1982 levels of world production of coal (2,700 million tonnes/year), reserves would last 190 years. Proven plus possible Soviet and other reserves would last between 3,000 and 4,000 years. Clearly there is no shortage of coal in the world in the immediate future.

Of the various features that affect the availability of coal reserves, two are especially relevant to the Third World: first, the very uneven distribution of "recoverable" reserves (see Table 5.4), and secondly the fact that most developing countries have very few if any coal reserves of any kind. The geological structure of the earth's crust is well enough explored for it to be known that coal (and lignite) is very unlikely to be discovered in many areas.

As they industrialized, some energy-deficient (now developed) countries (e.g. Italy and Japan) used imported coal; but these and others have since come to use other imported sources of energy, especially oil, and also hydroelectric and nuclear power (e.g. France recently). Most of the developing countries have had to industrialize without home supplies of coal and they have been able to import only small quantities. If China and South Africa, both well endowed with coal reserves, are excluded, 54% of the world's population in the rest of the developing world has only a few percent of the recoverable coal reserves.

Oil has been used commercially as a fuel for about 120 years and in large quantities for about half that time. Today it is the largest source of commercial energy in the world and is generally cheaper to extract, easier to transport and more versatile in its applications than coal, while being safer to transport by sea and more versatile than natural gas. In contrast to coal, most of the oil reserves are in the developing countries of the world, including the Middle East. According to the *British Petroleum statistical review of world energy* (BPSTAT; see Table 5.5), developing countries have over 80% of the world's 96,000 million tonnes of proved oil reserves (at the end of 1984), the Middle East alone having 56.4%. The members of the Organization of Petroleum Exporting Countries (OPEC) have more than two-thirds.

Two outstanding features of the world distribution of oil reserves that affect the Third World may be noted. First, the availability of oil reserves, like that of coal reserves, varies greatly from country to country (see Table 5.1, column 5), with many countries having virtually no oil and some not the least prospect of discoveries on their territories. Secondly, the 80% of the oil reserves found in developing countries are themselves highly concentrated. The fortunate citizens of Kuwait, Abu Dhabi, and Saudi Arabia count their oil reserves in thousands of tonnes per head, while in such major countries as China (3 tonnes), Brazil (2 tonnes) and India (1 tonne), reserves are minute per head of population.

The general picture of the world oil situation is that particularly before the sharp rises in oil prices in 1973 and again in the late 1970s the developing

Table 5.5 Reserves of oil and natural gas, by country.

	Oil					Natural gas	
Country	reserves (10^9 tonne)	life† (years)	population (10^6)	reserves/population (tonne/inhabitant)	Country	reserves (10^{12} m^2)	lif (ye
Saudi Arabia	23.0	★	11	2,090	USSR	41.1	70
Kuwait	12.4	★	2	6,200	Iran	13.6	★
USSR	8.6	14	278	31	USA	5.6	12
Mexico	6.8	45	80	85	Qatar	4.2	★
Iran	6.6	61	45	146	Saudi Arabia	3.5	★
Iraq	6.0	★	16	375	Algeria	3.1	★
USA	4.4	9	264	17	Canada	2.6	36
Abu Dhabi	4.0	★	1	4,000	Norway	2.5	86
Venezuela	3.7	38	17	218	Mexico	2.2	61
Libya	2.8	★	4	700	Netherlands	1.9	28
China	2.6	23	1,042	2	Venezuela	1.6	95
Nigeria	2.3	★	91	25	Malaysia	1.4	★
UK	1.8	14	56	32	Indonesia	1.1	62
Algeria	1.2	27	22	55	Nigeria	1.0	★
Indonesia	1.2	17	168	7	Kuwait	0.9	★
Canada	1.1	15	25	44	China	0.9	69
Norway	1.1	32	4	275			
World	96.1	34	4,845	20	World	96.2	60

Source: *BP statistical review of world energy*, various tables.
† Life of reserves in years at current rate of production (★ over 100).

countries that have oil reserves exported more oil than they consumed. Those like Saudi Arabia and Venezuela, with comparatively small economies, could not absorb large amounts at home but even so have been profoundly affected. Others, with large populations, especially Nigeria and Indonesia, could only derive local benefits from the oil consumed at home. The many developing countries without oil themselves could not afford to import it in more than small quantities, for example for essential needs such as public transportation or for luxury use such as private motoring. They continue to burn large quantities of fuelwood and even animal dung, both of which could be put to better use.

Natural gas has become a major source of fuel in the world even more recently than oil. Though it can usually be extracted very cheaply and moved by pipeline easily on land (e.g. in the USA and the USSR) it is not transported extensively by sea. Even more than with coal and oil, natural gas is highly concentrated in certain countries (see Table 5.5, column 6). Among developing countries Iran is especially well endowed.

Natural gas is without doubt useful locally in a number of developing countries, but the scale of reserves must be kept in proportion. For example the discovery of natural gas in fuel-deficient Bangladesh was hailed as a great boost for the country, yet in 1984 the total reserves were about 200×10^9 cubic meters, equivalent to about 3 tonnes of coal per head, the amount of energy consumed in about four months by the average US citizen.

There are many possibilities for discovering and using new sources of fuel. Oil is found in large quantities in tar sands (Venezuela and Canada) and shales (USA and USSR). It is thought by some (see Gold & Soter 1980) that there are vast reserves of "deep-earth gas" in the earth's crust, including methane of nonbiological origin, which could augment hydrocarbons of biological origin. Nuclear power, solar power and many other sources of energy exist. Without exception it is the present developed countries that have the capital, research facilities and technology needed to develop such sources. A very rough estimate of the numbers of scientists and engineers engaged in research and experimental development (*United Nations statistical yearbook 1981*, Table 70) credits the developing world, including China, with some 300,000 for 75% of the population of the world and the developed world (even after the Soviet and East European definitions have been watered down) with 2,500,000 for 25% of the population of the world.

Non-fuel minerals

In *Mineral facts and problems* (1980), some 80 different economic nonfuel minerals are covered. They range from lime and mica to diamonds, chromium and silver. For simplicity 15 have been chosen for consideration in this section, including those such as iron, aluminium, and copper, which have the greatest value of production; and also key metallic and nonmetallic minerals such as chromium, phosphates, and industrial diamonds. Although nonfuel mineral deposits are highly concentrated in certain parts of the world their distribution in total is very complex.

The data in Table 5.6 show that for 15 nonfuel minerals of particular importance, developed countries have the largest deposits of seven, developing countries (including South Africa) of eight. On the whole the developed countries of the world have been more thoroughly explored than the developing ones. Even so, it is remarkable that South Africa, Canada, and Australia, with only about 1.5% of the total population of the world, have between them over half of the known reserves of gold, platinum, chromium, manganese, and potash and over 20% of the world's iron ore, bauxite, lead, nickel, zinc, and asbestos. Europe and Japan, on the other hand, are very poorly endowed. If South Africa is excluded from the developing countries, these latter have less than half of all nonfuel mineral reserves, and China and India are particularly poorly endowed.

As with fossil fuels, a large proportion of the minerals extracted in developing countries goes to the developed countries. Some of the material returns in the form of manufactured products, much value usually having

Table 5.6 Largest reserves of 15 nonfuel minerals.

Mineral	Country	% of world total
1 Iron	USSR	30
2 Manganese	South Africa	53
3 Nickel	Australia	35
4 Chromium	South Africa	68
5 Aluminium	Guinea (West Africa)	29
6 Copper	Chile	20
7 Zinc	Canada	19
8 Lead	USA	21
9 Tin	Indonesia	15
	China	15
10 Silver	Canada	20
	USSR	20
11 Gold	South Africa	51
12 Potash	USSR	44
13 Phosphates	Morocco	53
14 Sulfur	Canada	14
	USSR	14
15 Industrial diamonds	Zaire	74

Source: *Mineral facts and problems* 1980, various tables.

been added. The prospects for the future are therefore as follows. The developing countries that export nonfuel minerals can stop or reduce exports in order to wait until, if ever, they become sufficiently highly industrialized to manufacture their own minerals rather than just preparing and processing them for export. On the other hand they can encourage the extraction of their minerals, often with foreign capital, almost invariably with foreign technical expertise, and export as much as possible in order to import capital goods (rather than others) to hasten their own industrialization.

It has been noted by Manners (1981) that the industrialized countries have tended to prefer investing in the mineral industries of certain countries such as Australia, Canada, and Brazil that are regarded as being stable politically, and to avoid unstable parts of the world. In this respect, southern Africa may be regarded as risky in view of the social unrest and the many local military activities in that part of the continent. As a result, not many major new discoveries of minerals have been made in the last two decades or so in areas such as the Andes, southern Africa and Southeast Asia, traditionally regarded as potential sources of minerals.

Conclusions

Many development studies are rightly concerned with local and regional problems round the world – an irrigation project here, local crafts there, squatter settlements somewhere else. In my view it is necessary at times to take a global view of development. Although the world is not organized politically or economically as a single unit one cannot escape the prospect, expressed by Mackinder (1904), that the world has become one system and that any event anywhere can have repercussions widely round the world.

The purpose of this chapter, then, has been to give some basic facts about the distribution of natural resources in the world. The reader may refer to Mesarovic & Pestel (1975), *Mankind at the turning point*, for an assessment of the future prospects for both agricultural production and nonrenewable natural resources, though the regional breakdown of information is very restricted. Much useful and interesting data is also available in Barney (1982). A particularly thought-provoking article by Keyfitz (1976) contains references to the role of natural resources. Attention is also drawn to Cole (1981), *The development gap*. Many people blame the unequal levels of development and living standards around the world on political constraints, such as the barriers to flows of people, goods, and information that are maintained round each sovereign state. I argue that even under a world government, with public ownership of all natural resources and means of production, and central plannning close perhaps to the Soviet model, it would still be physically impossible to move enough people or products or both quickly enough to modify greatly the present situation, at least for some decades. After all, the Soviet Union has been going for nearly 70 years now, yet Soviet geographers and planners still despair of creating homogeneous economic conditions throughout the industrialized West, resource-rich East, and developing South of their country. How much more difficult it would be for the whole world.

The gap between the rich and poor countries of the world seems to have changed little in recent decades. To what extent the uneven distribution of natural resources in the world have affected the situation is left to the reader to ponder over. It may help, however, if one imagines the complete cessation of all trade and other transactions between the developed and developing countries. Who would suffer (or benefit) more? With regard to bioclimatic resources in particular it is also perhaps worth asking whether if human beings were actually livestock, being raised on various farms by "super" farmers, they would be distributed according to sources of fodder the way they are at present. Are not some parts of the world greatly overstocked? Perhaps we humans are just another species of parasite which has, however, begun to exceed the capacity of those species, plant and animal, that support us.

References

Barney, G. O. (Study Director) 1982. *The Global 2000 report to the President*. The Technical Report, Table 6–13, 99. Penguin Books: Harmondsworth.

BP statistical review of world energy 1985. London: The British Petroleum Company.

Brown, L. R. 1978. The worldwide loss of cropland. *Worldwatch Paper* **24**. Washington, DC: Worldwatch Institute.

Cole, J. P. 1981. *The development gap*, Chichester: Wiley.

Eckholm, E. 1975. The other energy crisis: firewood. *Worldwatch Paper* **1**. Washington, DC: Worldwatch Institute.

Flower, A. R. 1978. World oil production. *Scientific American*, Vol. 238, **3**, March, 42–9.

FAO production yearbook 1983. Rome: Food and Agriculture Organization.

Gold, T. & S. Soter. The deep-earth-gas hypothesis. *Scientific American*, Vol. 242, **6**, June, 130–7.

Keyfitz, N. 1976. World resources and the world middle class. *Scientific American*, Vol. 235, **1**, July, 28–35.

Liu Zheng *et al*. 1981. *China's population: problems and prospects*. Beijing: New World Press.

Mackinder, Sir H. J. 1904. The geographical pivot of history. *Geographical Journal*, Vol. XXIII, **4**, April 1904, 421–2.

Manners, G. 1981. Our planet's resources. *Geographical Journal*, Vol. 147, 1–22.

Mesarovic, M. & E. Pestel 1975. *Mankind at the turning point*. London: Hutchinson.

Mineral facts and problems (1980 edn.). Bureau of Mines Bulletin **671**. Washington, DC: US Department of the Interior (Bureau of Mines).

Population Reference Bureau 1985. *1985 World population data sheet*. Washington, DC: Population Reference Bureau.

Simon, J. L. & H. Kahn 1984. *The resourceful earth: a response to Global 2000*. Oxford: Basil Blackwell.

United Nations statistical yearbook 1981. New York: United Nations.

6 *Development and resources in the Third World*

ROBERT S. BEDNARZ and JOHN R. GIARDINO

Introduction

When one examines the condition of less developed countries today, it is difficult to ignore the long list of problems which these nations face. Although it is possible to find examples of countries which have made significant progress in providing a better life for their citizens, it is also easy to identify many which have not. If asked to evaluate the quality of life in the Third World, most individuals living in developed nations would probably respond "bad!" But how does one evaluate the quality of life in a country? What attributes about a society can be used to measure this quality: gross national product (GNP), the amount of spendable income, the crime rate, the life expectancy, or social amenities? Numerous attempts to measure the welfare of a population have been made but most have met with little success. The purposes of this chapter are to discuss the methods that have been developed in order to evaluate and compare the quality of life among and between countries; to present a new quality-of-life index; to examine some of the problems associated with quality-of-life indexes; and to explore what connections, if any, exist between a nation's resource base and its quality of life.

Before discussing the indexes that have been developed to quantify the quality of life in a country, we should define what the concept "quality of life" means. One of the major problems in dealing with the concept of quality of life is that the concept varies from culture to culture. Factors important for one culture may be meaningless or unimportant in another culture. The quality of life which a group of people enjoy is obviously the product of a large number of variables, and there is little or no agreement among researchers about which variables should be included in a measure of welfare, nor about what should be the relative importance of the various factors. However the concept is defined, we think it should include at least the following components: economic and personal security, opportunities for personal and societal development, and adequate social services.

If cross-cultural comparisons are desired, the different tastes and preferences of the populations, the difficulty of converting wages and costs of other economic measures to a common currency base, and the inability to measure accurately the purchasing power of an amount of money from country to country add to the problems. Concerning the last two problems, the World Bank states: "For want of a better alternative, comparisons of income levels across countries have for years been made at official exchange rates. These comparisons yield estimates that are known to misrepresent the actual purchasing power of currencies. In particular, they tend to understate incomes in poor countries relative to those in rich ones" (World Bank 1982, p. 23).

Choosing a set of variables to measure welfare, let alone determining how to weight the variables, is extremely difficult. On the other hand, selecting different variables for each culture or adjusting the weights attached to them defeats the goal of developing a consistent measure which can be applied cross-culturally. To a large extent these problems occur because prices of goods traded worldwide tend to be quite similar from country to country. Unfortunately, many goods and most services are not traded. Wage differences from country to country reflect, in part, the differences in labor productivity in the production of traded goods. In much of the service sector, however, labor productivity varies little from country to country. In countries with high incomes and high wages services are expensive, while in poorer countries these same services are significantly cheaper. Not unexpectedly, then, because services are, for the most part, not traded worldwide, any measure that fails to take their lower cost in poorer countries into account will underestimate the buying power of wages in less developed poorer nations (World Bank 1982).

Two measures which have often been used to compare the wellbeing of populations across countries are per capita GNP and per capita GDP. Both of these indexes suffer from many of the problems just discussed. Neither measure takes into account the many factors that most would agree are important to the quality of life of a population, such as environmental quality, educational opportunity, access to medical facilities, or even the distribution of income within the society. According to the World Bank, however, the difficulty of incorporating these variables into an index is so great that often economists ". . . settle for partial measures such as GNP – which at least covers most of the goods and services available to meet important consumption needs" (World Bank 1982, p. 20).

Although a measure such as per capita GNP is unarguably an incomplete measure of quality of life, it should be pointed out that it is, nevertheless, an important statistic. In free-market economics, in which individuals are expected to provide most of their needs for themselves, the amount of money they have at their disposal is certainly very important. Even though per capita GNP does not directly measure many factors important to quality of life, it would not be surprising if variables associated with higher levels of welfare were directly associated with higher per capita GNP. If people have greater economic resources, they can create a market for medical facilities, educational opportunities, and the like. Even if this line of reasoning is sound and applies to a large number of societies, however, per capita GNP and other measures like it still suffer from a major problem. None of the measures takes the distribution of income within a country into account. Therefore per capita GNP may overstate the ability of an individual to purchase goods and services to improve his or her quality of life. It may also overstate the market for, and thus the availability of, these goods and services if a small segment of the population earns a large portion of the country's income.

Quality-of-life index

Some researchers (e.g., Morris 1979, Gonzalez 1983) have tried to overcome some of these problems by relying on a different type of measure, the Physical Quality-of-Life Index (PQLI). As the name suggests, this measure concentrates on physical characteristics of the population group to provide a measure of the quality of life. The most common formulation includes three factors: infant mortality rate, life expectancy, and literacy rate. The PQLI ranges in value from 0 to 100 and is the average of the scores for each of the three variables. The highest value found for each of the variables is assigned a value of 100, and all lower values are expressed as a percentage of the largest. Gonzalez' 1983 index included more than three measures, but the variables he selects are very similar to the usual three, and he handles them in essentially the same fashion. Not surprisingly, Gonzalez' ranking of world nations looks very similar to that generated by applying the PQLI to the same countries. It can be argued that Gonzalez' index is more comprehensive and sophisticated than other measures, however, and it should be noted that the index does make useful distinctions between countries that might otherwise be lumped together.

Development of natural resources

The rôle that resources play in the direct development of a country is difficult to evaluate for several reasons. one of the major difficulties is that the value of resources cannot be accurately measured and evaluated. In addition, the total Third World resource base is incompletely understood with respect to soils, climate, vegetation, water supplies, and minerals. Further, the interdependence between cultural and physical factors makes it difficult to evaluate the true importance of any specific variable. Considered in this light, development of Third World countries is probably a function of the cultural systems operating on the available physical resource bases.

Some geographers have attributed too much importance to physical resources as compared to the available cultural base. Natural resources are not responsible for development and economic growth, as is supported by the low correlation between resource base and degree of present development. For example, Cole (1981) pointed out that there are many highly developed areas with poor resource bases such as the northeastern USA, Japan, and the UK. On the other hand, Indonesia, Zimbabwe, and Venezuela, with better resource bases, lag in economic development.

At the other end of the spectrum, economists often prefer the notion that economic development is not at all dependent upon internal resources, but is instead controlled by the skills, by labor, and by a technology applied to either domestic or imported raw materials. Many countries with high resource potentials are restricted by external trading potential, capital accumulation, labor availability, or population skills and technology.

A similar line of reasoning underlies the following quote by Epstein (1985) in the *New York Times*. In it the author not only de-emphasizes the importance of natural resources as a cause for economic development, but goes on to include the capital stock of a nation:

> Imagine, for example, that it is 1785 in the United States. Our capital stock includes such things as wooden plows, draft animals, hand pumps and Franklin stoves. Now perform a mental experiment on this economy: Assume that the rate of capital accumulation quadruples – but that innovative activity stops permanently. There would be some improvement in material well-being, but nowhere near what actually happened. Now try a mental experiment from real life – Germany and Japan at the end of World War II. Their capital stock was about comparable to that of many undeveloped countries. But twenty years after the war ended these countries had achieved levels of prosperity that far exceeded their prewar levels. The inherent advantage of Germany and Japan was the "knowledge base" embodied in their people; this proved to be more decisive than a capital base (Epstein 1985, p. 2).

The truth probably lies somewhere between the views of the economists and some geographers. Resources are often important, especially during early stages of development, when economic activity often amounts to no more than the exportation of raw materials. In areas of very limited human skills and applicable technology, the abundance of physical resources is essential. When population skills, labor, technology, capital, and trading potential are high, the internal physical resource base plays a much less important rôle. Often, in more developed countries, diversification of resources becomes a major economic influence.

Barke & O'Hare (1984) also make this point when they emphasize that the importance of natural resources varies with the stage or level of development. They point out that in developed nations a lack of resources can be compensated for by substituting other factors of production (e.g., skilled labor or capital). Less developed nations, however, often lack these other factors and must therefore rely more on what they do have – natural resources.

Resource exploitation is constrained by the value, the concentration and the location of economic raw materials and the transportation system available. The poor distribution of resources or the inability to move them to where they are needed can be the controlling limitation for economic development. This contention is supported by the situation of many underdeveloped, Third World countries. On the other hand, the destruction and depletion of renewable resources and nonrenewable resources in many developed nations suggests that less-developed resource exporters may have an opportunity to use the sale of their resources to start an economic development program. The danger is that a level of development conducive to diversification may not be reached prior to resource exhaustion in the developing country.

Regardless to which view of the importance of resources one ascribes, it is important to note that recently researchers have begun to pay much closer attention to agriculture and agricultural resources. In part, this renewed interest in agriculture stems from the failure of past development plans. These programs promoted industrialization and the development of mineral resources and often directly or indirectly resulted in increased urbanization. Although not all planners agree about the importance of agriculture, interest appears to be building, as the following statement by Kristof illustrates. "Today the watchword of development is not industrialization, as it was in the 1950s, 60s and 70s, but agriculture" (Kristof 1985, p. 1).

The reasons for the failure of industrial development plans are many and are often difficult to identify precisely. To make matters worse, the initial failure often fed the vicious cycle of poverty operating in many of the poor nations. To quote Kristof again:

> The roots of the failure of heavy industrialization are varied and complex. Local workers often lacked the training to run sophisticated factories and machines. Countries ran out of cash to pay for spare parts and other imports needed for production. Industries and urban workers were coddled with subsidized food and jacked-up exchange rates intended to keep imports cheap. But low food prices discouraged farmers from producing and countries found themselves allocating scarce foreign exchange to import the most basic of goods: food. As a result, many economies stagnated or grew very slowly. The cycle of poverty continued. (Kristof 1985, p. 8.)

Unfortunately, indirect problems caused by the failure of industrial development plans also surfaced.

> Industrialization had lured rural residents to cities, but they did not find the break from poverty they expected. Urban populations ballooned, but all too often, the former peasants found themselves in festering shanty towns separated by open sewers from the gleaming buildings and pin stripes of the business districts. (Kristof 1985, p. 8.)

Natural resources

While we accept the importance of the mineral resource base in providing a basic requirement for advancement towards industrialization, we find it very difficult to use mineral resources as a predictor of economic wellbeing. In this paper, we have not discussed mineral resources specifically. Our reasons for excluding minerals are based upon several criteria: (a) the known resource base is not finite – new discoveries continue to change the worth of mineral bases; (b) the data on mineral resources are sketchy and difficult or impossible to compare from country to country because of different reporting criteria; (c) the value of minerals is determined almost totally by external political controls (e.g., the Zambian experience with copper during the 1970s).

In the light of the importance of agriculture to developing countries now and in the future, it would seem advisable to undertake some discussion of those resources that have a bearing on agricultural potential and productivity. What follows is an attempt to evaluate briefly those agricultural resources and to point out some of the limitations those resources (or lack of them) place on developing countries.

CLIMATIC RESOURCES

Climate imposes restrictions on soils, vegetation, and animals through the availability of usable moisture and energy. These ecological restrictions control agricultural output and consequently often have an impact on the health of the population. Because the Third World comprises such a large area, great diversities of climate and physiographic variation are found. Underdeveloped countries seem to have in common abundant insolation; in only a few are biological potentials limited by a lack of solar energy or resultant low temperatures. In fact, portions of China, Taiwan, India, and the Philippines, for example, receive enough insolation to allow several successive crops to be grown annually, and therefore have potentially high agricultural yields. This same characteristic, however, severely limits soil fertility because of the accelerated depletion of organic matter associated with high temperatures (Biswas 1979).

Precipitation variations are much greater than variations of temperature in Third World countries. For the Third World as a whole, the availability of adequate water is the primary agricultural constraint. Just as high temperatures accelerate soil degradation, high precipitation also severely accelerates degradation by increased leaching, erosion, and more rapid decay of organic materials. Although multiple crops can be grown on these wet tropical and tropical monsoonal soils, the precipitation levels encountered favor tree and root crops over cereals and grains. In many of the tropical monsoon areas, low water availability can be a severe restriction on agriculture during dry seasons while excessive rainfall limits wet-season productivity. Where modern cropping systems are not utilized to protect the soil surface from high precipitation rates, severe damage often results as surface compaction, decreased infiltration, and increased runoff cause accelerated rates of erosion. The high precipitation variability and unpredictability are a primary climatic hazard and deterrent to agricultural production in underdeveloped Third World countries (FAO 1981). The recent drought in the Sahel serves as an example of the disastrous effects of a climate hazard.

VEGETATION RESOURCES

Theoretically, the vegetation distribution within Third World countries is controlled primarily by climatic limitations. Upon closer examination, it becomes evident that vegetation and potential agricultural production are limited by climate, soil, and relief. Agricultural production is further constrained by the availability of labor near potentially productive areas.

Forests are second only to soils in their control over the biotic resources of an area. They reduce surface damage from extremes of moisture availability, temperatures, runoff-induced erosion, and they stabilize the biomass of entire regions. In addition to the economic importance of forests, they provide habitat for many of the world's species of plants and animals. Economic gains are being made in some Third World countries from timber harvesting which was previously prohibitively costly. Because mechanized harvesting and higher prices have made timber exploitation possible and profitable, uncontrolled harvesting may result in depletion of this currently renewable resource. In those countries where even fuel for cooking and heating is scarce, local forest depletion would be particularly devastating (Jackson 1977).

Although the native domestic livestock and crops of an area are usually suited to poor soil quality and local climates, they are often ineffective food producers. Current attempts to establish high-yield varieties have met with low success rates. In contrast to domestic livestock, natural wildlife varieties are commonly very efficient meat producers. Failure to establish production and marketing of these animals can be attributed to acquired (or unacquired) tastes and to handling, transport, and marketing difficulties, some of which may be overcome with adequate economic incentives. For example, the government of Zambia initiated a program in 1972 to market elephant meat, yet after five years the program had achieved little success.

Biological losses of crops and livestock are largely attributable to conditions conducive to the widespread growth of pests and parasites. Warm moist tropical environments favor rapid-growing forbs – competitors with crops for nutrients and moisture – and many parasites, including worms, and viral diseases which damage crops, injure livestock, and plague the human population. Many particularly damaging pests are able to survive in these tropical and subtropical countries because of the lack of killing frosts or freezes. The locational limitations of these diseases inevitably means that they inflict more harm on the poorer countries than on the developed countries located in more moderate climates.

SOIL RESOURCES

Although numerous soil types are found throughout the Third World, most think these soils are inferior to those of developed countries. This opinion may be largely unwarranted. The most common soil orders found in these regions are the subtropical ultisols and tropical oxisols (Steila 1976). These similar soil types vary in degree of weathering, silica, and nutrient content: but all are characterized by high leaching, low organic matter, poor structure, and resultant low effective porosity. Oxisols of tropical forests are usually highly leached, acidic, and very low in nutrients. Oxisols and ultisols of the savanna grasslands are commonly less intensely leached and contain higher concentrations of organic matter and nutrients. Although these soils seem much better suited for agriculture, the latter portion of the growing season is often subject to drought conditions, and thus the hard, desiccated soil cannot be used effectively for crops. On the other hand, the ultisols of arid and semi-arid regions are seldom leached and usually have a very high nutrient content. With adequate irrigation these lands can produce high crop yields, but long-term water deficiencies usually prevent irrigation.

The controlling soil factors presented so far suggest a poor potential for agricultural production in Third World countries. This is not the case everywhere. In many areas in eastern and southeastern Asia, large expanses of fertile alluvial soils have been exploited for food production. In Latin America fertile river valleys and deltas also produce high crop yields. In other localities, recently developed volcanic soils provide abundant food and market crops. It is evident that agricultural development is dependent upon many variables: soil fertility, climate, pests and diseases, and management practices. Improvements in agricultural techniques are vitally important, but they must be compatible with the controlling conditions of an area.

Data

In order to explore the connection between the resource base of a country and the quality of life of its population, a data set consisting of 38 nations was collected. Countries were selected to cover a wide range of standards of

Table 6.1 Lists of variables used to evaluate quality of life and resource base.

Resources*	Quality of life†
Aluminium reserves	Income distribution
Lead reserves	Per capita GNP
Coal reserves	Defense expenditures
Cobalt reserves	Child mortality rate
Copper reserves	Secondary-school attendance rate
Gold reserves	Infant mortality rate
Iron reserves	Calories consumed per capita
Manganese reserves	Life expectancy
Natural-gas reserves	Literacy rate
Nickel reserves	
Petroleum reserves	
Silver reserves	
Titanium reserves	
Vanadium reserves	
Zinc reserves	
Harvestable timber	
Arable land	
Grain production	
Reserves of other precious metals combined (e.g., platinum)	

Sources:
* Bureau of Mines 1980. *Mineral Facts and Problems, Bulletin 671*. Washington, DC: US Department of the Interior.
* Bureau of Mines 1984. *Minerals Yearbook, Vol. 3, Area Reports, International*. Washington, DC: US Department of the Interior.
† The World Bank 1982. *World Development Report 1982*. New York: Oxford University Press.

living and resource endowments. Some concessions were made with regard to data availability. The latter was especially important with respect to income-distribution statistics. Unfortunately, this information is accessible for fewer countries than we expected.

Nevertheless, it was hoped that adding income-distribution information to a measure such as per capita GNP would greatly improve its ability to capture important characteristics of a population's quality of life. After all, measures like per capita GNP have been used with fairly good results by other researchers even though they do suffer from the shortcomings previously discussed. By gaining a more accurate picture of the economic resources of the average citizen of a country (by considering income distribution), it seemed likely that the measure's performance could be greatly improved.

Thirty-two variables were collected for each of the countries (see Table 6.1). Among these were 20 that pertained to the countries' resource bases. These included figures measuring precious metals, energy resources, and agricultural resources. Included among the quality-of-life variables were economic measures such as per capita GNP, health-related factors such as mortality rates and dietary variables, and social factors such as literacy and school-attendance rates. Several indexes were also constructed and variables were often transformed to per capita figures. In general, all the available data which the scope of this study allowed were collected and processed.

Analysis

To determine the impact income distribution might have on economic factors, World Bank income-distribution statistics were used to estimate per capita GNP for the poorest 80% and 90% of the populations of the countries in the sample. Figures for the lowest-earning 80% and 90% were used in order to eliminate the bias introduced into the per capita figures by the small proportions of the population with very high incomes. This problem is most severe in poor, less developed countries. Table 6.2 shows the results of these calculations.

The first column of the table is a ranking of 38 countries by per capita GNP. The second column shows the ranking of the same countries by per capita GNP for the lowest earning 80% of the population. The proportion of the total GNP allocated to the 80% of the population is the same as the fraction of the total income of the country that this segment of the population earns. In other words, if the bottom 80% of the population earns 60% of the nation's total income, then 60% of the country's GNP is divided by a figure equal to 80% of the nation's population in order to calculate the per capita GNP figure. The third column, which shows the countries ranked by per capita GNP for the bottom 90% of the population, was calculated in a similar manner.

As can be seen from the rankings, taking distribution of income into account does little to change the ranks of the countries. Only six countries move more than two ranks in the second column and only four countries in the third column. Furthermore, half of these rank changes occur beyond rank 27, so that including distribution-of-income statistics in a per capita GNP measure seems to have little impact on the results, especially for the poorer countries. Statistical measures of the similarity between the raw GNP figures and those adjusted to take distribution of income into account also support this conclusion. Both Spearman and Pearson correlation coefficients were calculated for the adjusted and raw per capita GNP figures, and, in every case, the correlation was greater than or equal to 0.99. Scatter plots of the data also showed the same result, with the plotted points rising from the origin in a nearly perfect straight line. The vast differences in the raw data and the systematic change in income distribution as income and wealth increase yield adjusted figures which are remarkably similar to the untransformed data.

Because the addition of income distribution had little effect on per capita GNP, a PQLI-style index was developed. Like the standard PQLI, three variables

Table 6.2 Countries ranked by per capita GNP.

Ranked by per capita GNP		Ranked by per capita GNP of the lowest 80% of the population		Ranked by per capita GNP of the lowest 90% of the population	
1 Bangladesh	130	Nepal	71	Nepal	83
2 Nepal	140	Bangladesh	94	Bangladesh	105
3 Malawi	230	Malawi	142	Malawi	153
4 India	240	India	152	India	177
5 Sri Lanka	270	Tanzania	174	Tanzania	200
6 Tanzania	280	Sri Lanka	191	Sri Lanka	215
7 Kenya	420	Kenya	208	Kenya	253
8 Indonesia	430	Honduras	225	Honduras	311
9 Honduras	560	Indonesia	272	Indonesia	315
10 Philippines	690	Philippines	397	Philippines	472
11 Peru	930	Peru	453	Peru	590
12 Turkey	1,470	Turkey	799	Turkey	969
13 South Korea	1,520	Panama	826	Panama	1,073
14 Malaysia	1,620	Brazil	856	Malaysia	1,084
15 Costa Rica	1,730	Malaysia	889	Brazil	1,125
16 Panama	1,730	Costa Rica	978	Costa Rica	1,163
17 Brazil	2,050	South Korea	1,039	South Korea	1,224
18 Mexico	2,090	Mexico	1,105	Mexico	1,379
19 Chile	2,150	Chile	1,306	Chile	1,558
20 Argentina	2,390	Argentina	1,485	Argentina	1,721
21 Yugoslavia	2,620	Yugoslavia	2,008	Yugoslavia	2,245
22 Venezuela	3,630	Venezuela	2,087	Venezuela	2,593
23 Hong Kong	4,240	Trinidad/Tobago	2,731	Hong Kong	3,237
24 Trinidad/Tobago	4,370	Hong Kong	2,809	Trinidad/Tobago	3,312
25 Spain	5,400	Spain	3,902	Spain	4,398
26 Italy	6,480	Italy	4,544	Italy	5,177
27 UK	7,920	UK	6,019	UK	6,706
28 Finland	9,720	USA	7,270	Japan	8,000
29 Australia	9,820	Japan	7,294	Canada	8,228
30 Japan	9,890	Canada	7,344	Australia	8,325
31 Canada	10,130	Australia	7,512	Finland	8,510
32 USA	11,360	France	7,047	France	9,058
33 Netherlands	11,470	Finland	8,894	USA	9,265
34 France	11,730	Netherlands	9,033	Netherlands	9,928
35 Norway	12,650	W. Germany	9,377	W. Germany	10,751
36 Denmark	12,950	Norway	9,914	Norway	10,935
37 Sweden	13,520	Denmark	10,117	Denmark	11,838
38 W. Germany	13,590	Sweden	10,613	Sweden	11,838

Source: World Bank 1982.

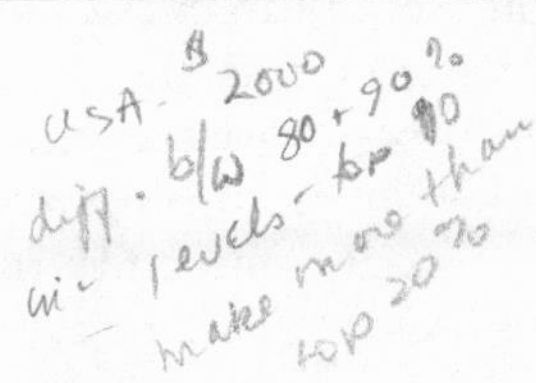

were converted to percentages of the best ranking scores and then averaged to yield an index with a theoretical range of 0 to 100.

We reasoned that the three most important factors related to quality of life involved the health, economic resources, and education of the population. Because infant mortality has such a significant impact on life span we thought the usual PQLI formulation suffered from a type of double counting by including both of these factors. We opted instead for childhood mortality rate as our single measure of the population's health. It was thought that childhood rather than infant mortality was a better measure of the whole population's health and was less dependent on access to a physician. To measure the population's economic condition, per capita GNP was selected. As argued earlier, this measure seems to be related to a surprising number of important quality-of-life factors.

Finally, as a measure of education, the percentage of the eligible population attending secondary school was chosen. This variable seemed preferable to literacy rate, which often varies little for nations in very different stages of development (e.g., Tanzania 66% and Indonesia 62%).

The rankings of nations produced using this index are similar to those yielded by Gonzalez' index (1983), and our index, like his, adds an economic factor to the usual PQLI formulation.

Despite working and reworking the quality of life factors, we could find little or no relation between the quality of life and the resource base. In an initial attempt to determine if a relationship existed, scatterplots of the quality-of-life variables and indexes versus the resource measures and indexes were produced. Unfortunately, the results were disappointing. Most of the plots showed little or no identifiable pattern, and many were dominated by large clusters of points near the resource axis. This phenomenon was the result of the wide range of resource endowments and the near or total lack of numerous resources on the part of many nations.

Because the initial attempts to establish a connection between resources and quality of life produced no significant results, the use of other statistical techniques was considered. And because our goal was to establish a relationship between the two variable categories, canonical analysis was selected as the most appropriate technique. For a detailed discussion of canonical analysis see Tatsuoka (1971). As a member of the set of multivariate linear statistical techniques, canonical analysis assumes interval data and linearity among variables. Thus, the technique identifies the elements of one set of variables that are most related, in a linear sense, to the elements of another set of variables (Nie *et al.* 1975).

We hoped that the canonical correlation would identify a group of resource-base variables that would account for a large proportion of the variation in the quality-of-life variables. Canonical correlation does not assume a cause-and-effect relationship. In fact, the direction of the relationship is unimportant, not unlike simple linear correlation. Unfortunately, the technique generated results which were no better than those of the other statistical techniques applied to these data. At the root of the problem is the very nature of the data set itself. Few of the variables examined show much of a relation to other variables in the data set. This was true of variables within the quality-of-life and resource-base groups as well as between the groups. Therefore, the canonical correlation analysis provided little insight into what connection, if any, exists between resources and quality of life.

The lack of relationships in the data set which the analysis revealed was not too surprising. After all, the other procedures had been unable to identify any simple relationships or correlations between pairs of variables or variable indexes. Nevertheless, the absence of simple relationships does not necessarily preclude the existence of more complicated multivariate correlations, and it was hoped that the canonical correlation would reveal something of this sort. In this case, however, the procedure proved fruitless regardless of the variables employed or the structure of the statistical model.

The results of the canonical correlation procedure did confirm the basic good sense displayed in the choice of variables for the index of quality of life. The variables we chose, childhood mortality rate, the percentage of the eligible population enrolled in secondary school, and per capita GNP, were identified by the procedure as three distinct components of the quality of life. Because our index is very similar to Gonzalez', the same would undoubtedly be true for the variables he selected. Thus there is good evidence that the indexes are composed of variables which do not overlap or serve as surrogates for each other. Each of the variables contributes at least somewhat uniquely to the quality-of-life index.

Besides the obvious results that follow from the data analysis, at least one other comment seems worth making. When setting out to explore the connection (or lack thereof) between a nation's resource base and its quality of life, we expected that some problems would be encountered regarding data availability, especially with respect to quality-of-life measures for less developed countries. Quite surprisingly, quality-of-life variables were relatively easy to collect compared with those required to characterize accurately a country's resource base. Many sources of these data aggregated minerals into less useful groups without reporting the ungrouped data. Others aggregated countries into regional or economic groups, which also lessened the data's usefulness. These problems were most severe for the smallest and least developed nations.

Conclusions

First, and most important, we must conclude that it is extremely difficult to demonstrate that there exists a significant connection between a country's resource base and its quality of life. Although there may be a number of reasons why it is difficult to establish this connection (e.g., data inaccuracies), we feel that the lack of a statistical relationship is not misleading but is instead an accurate representation of real world patterns. Having said this, we do not deny that resources can and do play an important rôle in the development and thus the quality of life in specific instances.

In any case, the importance of a country's resource base may be becoming a moot point. Old development theories and models are undergoing close scrutiny and ideas are changing rapidly. As has been noted, agriculture is receiving new attention from many development planners, and there has been a new emphasis on equity in the development process. Early development theory was based on the postwar reconstruction of Europe which ". . . suggested that investment capital was the limiting constraint to economic growth and processes of capital generation for investment in industry became the *sine qua non* of development strategy during the early period. The careless equated industrialization with development" (Eckert 1983, p. 36).

Strategies based on this line of reasoning often led to policies that concentrated the benefits of the program within a small group of the population which was already affluent. These policies were justified because "Economic theory suggests that the marginal rate of savings rises with incomes. Therefore, concentrating wealth and incomes among a limited number of recipients was seen as consistent with the need to accelerate economic growth through industrialization" (Eckert 1983, p. 36). In fact, Nobel Laureate W. Arthur Lewis went so far as to state "Economic growth is dependent upon the inequality of income. That this dependence exists cannot be denied . . . " (Eckert 1983, p. 39).

Despite the theoretical justification for the strategies, their results were not impressive. Furthermore, in those places where development did occur, large portions of the population remained impoverished. Alain de Janvry makes the observation that although economic growth and development has been strong in many parts of Latin America, it has not produced a more even distribution of income (de Janvry 1983).

Others have also questioned development strategies which do not include the rôle played by today's powerful multinational corporations. For example, Forbes (1984) argues that free access to the economies of underdeveloped nations by multinationals usually results in disaster. Often the nascent industrial or handicraft sector is destroyed as a result of the competition from outside. Thus, the sector of the economy that is probably necessary to serve as a base for modern capitalist development disappears.

In conclusion, a developing country's resource base seems significant only if the nation can use its resources in order to build a stronger, more developed economy. It is unlikely that the development process will be initiated or sustained if the nation simply exports its raw materials. Few places have significantly improved the quality of life of their populations by exporting resources. The wild swings in basic commodity prices are well documented. Booms are inevitably followed by busts. As the recent change in oil prices demonstrates, the challenge is to use the revenue generated by raw-material exports to further economic development. Relying on resource exports to raise the quality of life in a region over the long haul is a gamble that will almost surely be lost.

References

Barke, M. & G. O'Hare 1984. *The Third World*. Edinburgh: Oliver & Boyd.

Biswas, A. K. 1979. Climate and economic development. *The Ecologist* **6**, 188–96.

Cole, J. P. 1981. *The development gap: a spatial analysis of world poverty and inequality*. New York: Wiley.

de Janvry, A. 1983. Growth and equity: a strategy for reconciliation. In *Issues in Third World development*, K. C. Noble & R. K. Sampath (eds.), 19–33. Boulder, Colorado: Westview Press.

Eckert, J. B. 1983. Income distribution and development: a critique of current methodology. In *Issues in Third World development*, K. C. Noble & R. K. Sampath (eds). Boulder, Colorado: Westview.

Epstein, E. 1985. End the drag of special tax breaks. *The New York Times*, August 25, Section 3.2.

FAO 1981 *State of food and agriculture, 1980*. New York: Food and Agriculture Organization.

Forbes, D. K. 1984. *The geography of underdevelopment*. Baltimore: The Johns Hopkins University Press.

Gonzalez, A. 1983. Major world countries: levels of socioeconomic development. Paper presented at Denver: Annual Meeting Association of American Geographers, April 1983.

Jackson, I. J. 1977. *Climate, water and agriculture in the tropics*. London: Longman.

Kristof, N. D. 1985. The Third World: back to the farm. *The New York Times*, July 28, Section 3.1.

Morris, M. D. 1979. *Measuring the condition of the world's poor: the physical quality of life index*. New York: Pergamon & Overseas Development Council.

Nie, N. *et al*. 1975. *Statistical package for the social sciences*, 2nd edn. New York: McGraw-Hill.

Steila, D. 1976. *The geography of soils*. Englewood: Prentice-Hall.

Tatsuoka, M. M. 1971. *Multivariate analysis*. New York: Wiley.

World Bank, The, 1982. *World development report 1982*. New York: Oxford University Press.

7 *Defining and identifying NICs*

GERALD L. INGALLS and WALTER E. MARTIN

Introduction

In search of simplified explanations for the differences in levels of economic development among nations, it is often convenient to resort to dichotomous measures like rich/poor, have/havenots, developed/underdeveloped, or even us/them. Rarely are such measures sufficient. Often they serve little else but to reinforce existing prejudices. What is needed is orderly classification by which specific nations can be systematically and objectively compared. Such classification systems are useful devices for ordering into structured and comparable subsets the tremendous volume of what is often very complex information available on the economic, social, and demographic welfare of countries.

Extremes in economic classification systems are reasonably clear-cut and easy to characterize. On the one hand there is the US; on the other there is Haiti. It is not normally the extremes that present problems in categorization, but the blurred boundaries between classes. Such cases frequently offer the best opportunities to grasp the principles and concepts on which classification systems are predicated. A case in point is one of the subsets of many economic classification systems – the "newly industrializing country" or NIC.

Focusing on the pedagogical issue of effectively utilizing the development continuum, consider the dilemma posed by categories of states like NICs. Without clearer measures of the boundaries which separate NICs from either the other industrialized countries or from less industrialized countries, can the factors which govern progress through the developmental sequence be fully appreciated? The major goal of presenting such classification systems is to develop the economic, cultural, and geographic principles which underlie international, national, and subnational patterns of economic development. NICs can serve as valuable case studies which allow instructor and student to present, develop, and debate these principles and to address future patterns of economic development.

In this essay we develop an operational definition of an NIC: that is, one that can be used to position NICs within an economic classification system. Using the characteristics specified within this definition, we employ a two-step procedure to identify a set of NICs. Finally, we compare the list of NICs identical in this way to NICs derived in other recent publications, and provide a brief economic description of each country identified as an NIC.

Defining NICS

Like most categories or classes of countries lying between the extremes of the developmental classification system, NICs are in transition from one class to another. Most definitions suggest, either by implication or by specific threshold criteria, that NICs are positioned within the categories of states that have middle levels of per capita Gross National Product (GNP; Forbes 1984), which is frequently the basis for measuring the economic welfare of a nation. Although NICs have been variously, and, all too often, ambiguously defined, one of the initial characteristics prevalent in most definitions is the idea that NICs are developing nations (Cody, Hughes & Wall 1980, Crow & Thomas 1983, Dadzie 1980, Forbes 1984) which have not yet quite achieved the economic status or level of per capita GNP of the wealthy, industrialized societies.

Concomitant with the characterization of NICs as middle-income and developing is the idea that NICs are states whose economies have nascent developed industrial sectors. NICs are states that have moved through the take-off stage of economic development quite recently relative to countries with established industrial sectors (Reynolds 1985). They have demonstrated a rapid and sustained economic growth, particularly industrial growth, through the decades of the 1960s and 1970s (Cody, Hughes, & Wall 1980, Crow & Thomas 1983, Dadzie 1980, Forbes 1984). A report published by the Organization for Economic Cooperation and Development (OECD) in 1979 characterized NICs by the rapid expansion in the "level and share of industrial employment," by steady growth in the level of manufacturing exports, and by the rapid decrease in the gap between the per capita income in NICs and that in advanced industrial countries (Crow & Thomas 1983).

On this latter point of narrowing the gap, Rostow concurs. He suggests that the US narrowed the industrial and developmental gap with Great Britain; Sweden, Japan, Russia, Canada, and Italy closed the gap with the US and western Europe; and so will a number of countries which have moved through the take-off stage only during the second and third quarters of this century close the gap with the more established industrial countries (Rostow 1978).

Kuznets (Jumper *et al.* 1980) avoided the use of distinctive stages of growth developed by Rostow, leaning instead to broader "zones" of development or transition. He suggested that modern economic development be distinguished from non-modern or traditional types of growth, and he provided several characteristics of modern economic development, including:

(a) minor increase in the per capita national production sustained over a period of at least two or three decades;
(b) major shifts in the industrial structure of product and labor force;
(c) major shifts in the population, i.e., urbanization; and,
(d) a decreasing reliance on agriculture.

These characteristics appear remarkably similar to Rostow's characterization of the changes that characterize a transition from a traditional to a mass-consumptive society. While Kuznets disdained the use of the term "take-off" as too rigid a delimiter from what had gone on before, and preferred instead to speak of phases of early modern economic growth there is, at least on these points, similarity to Rostow's take-off stage and the early phases of drive to maturity.

While they may disagree on the criteria used to distinguish one phase of development from another, in general such concepts of economic development have the common base of progression from one phase to another. As Kuznets (Jumper *et al.* 1980) points out, this developmental progression may not be constantly in the same direction. There can be economic highs and lows. But once set in motion the progression is relatively continuous. Some countries, such as NICs, make the transition from one end of the scale to the other more swiftly than others. The speed with which economic and particularly industrial development occurs is an important element in most conceptual outlines of an NIC.

It is apparent that any definition of an NIC must be multi-faceted consisting of several primary elements. First, an NIC is likely to be a developing economy with middle-range per capita (GNP) income levels somewhat less than that of more established industrial economies but considerably greater than that of other developing countries. Second, NICs are likely to be rapidly industrializing or perhaps even fully industrialized nations (Crow & Thomas 1983). Industrialization implies not only an increase in the proportion of the Gross Domestic Product (GDP) devoted to the industrial sector (Crow & Thomas 1983) but also an increasing development of manufacturing specifically. New, more refined, and often technologically more complex, industries develop (Reynolds 1985). Third, the share of the newly developed manufacturing sector in total merchandise exports increases as new markets develop. Fourth, NICs are characterized by rapid economic and industrial growth which is sustained over some minimum period of no less than one decade and perhaps as much as two decades. Finally, the sustained economic growth of NICs has fueled significant increases in per capita income levels which has led to a closing of the gap between them and wealthier countries.

Figure 7.1 Commonly used developmental continua.

Generalized versions of developmental continua

Developing World
Poor World
Non-industrialized World
Traditional World

Developed World
Rich World
Industrialized World
Modern World

Fourth World — Third World — First/Second Worlds

W. Rostow's five stages of growth

Traditional — Precondition to take-off — Take-off — Drive to maturity — High Mass consumption

NICs on one version of a developmental continuum

Lesser Developed Countries — Resource Wealthy Countries — Newly Industrialized Countries — Wealthy Industrialized Countries

Identifying the NICs

Conceptually, how would countries identified as NICs fit within the developmental continua offered by Rostow or Kuznets? They are placed somewhere nearer the higher ranges of economic development (Fig. 7.1). Since most continua are based on economic criteria, countries are positioned in terms of wealth, which is normally measured in terms of national productivity such as GNP and is often standardized for population with a measure such as per capita GNP. As upper-middle-income countries, NICs are positioned nearer wealthier industrialized states at one extreme of the continuum towards which, Rostow and Kuznets concur, they are rapidly moving. Operationally, identifying specific NICs invokes less agreement.

It requires only a cursory search to discover that there is no list of NICs accepted by everyone. Definitions vary and so do the nations that are identified as NICs. The several countries frequently identified as NICs are: Brazil, Greece, Hong Kong (although not a genuine "country"), Malaysia, Mexico, Portugal, Singapore, South Korea, Spain, Taiwan, and Yugoslavia (Crow & Thomas 1983, Dadzie 1980). All of these states have been identified

Table 7.1 Defining NICs: characteristics and operational vehicles.

Characteristic of NIC	Literature source	Operation variable
Variables used in Step 1:		
Middle-income	1, 2, 3, 4	Per capita GNP, 1983
Industrialization:		
Level of employment	1, 2, 3, 4, 5, 6	Percentage of labor force in industry, 1981
Share of industry	1	Percentage of total GDP in manufacturing, 1983
Variables used in Step 2:		
These are variables which depict rapid rates of economic growth:	7, 1, 2, 3, 4, 5, 8, 9, 10	
Increase in percentage of GDP in industrial sector	1, 2, 3, 5, 6	Average annual growth rate of manufacturing as a percentage of total GDP, 1973–83
Change in structure of production	5, 6	Percent change in the percentage of GDP in manufacturing, 1965–83
Growth in the level of industrial employment	1, 2, 3, 5, 6	Percent change in the percentage of labor force in industry, 1965–81
Enlargement of export shares in manufacturing	1	Percent change in the percentage share merchandise exports in manufacturing, 1965–82
Rapid relative reduction in per capita income gap	1, 8	Percent change in the percentage difference in per capita income GNP between the US and each potential NIC

Sources:
1 Crow & Thomas 1983; 2 Dadzie 1980; 3 Forbes 1984; 4 Ingalls & Martin 1983; 5 Jumper *et al.* 1980; 6 Reynolds 1985; 7 Cody *et al.* 1980; 8 Rostow 1978.

as having passed through the take-off stage and are currently within what Rostow describes as the drive to maturity. Other sources suggest that some countries may have attained fully industrialized status – including Greece, Portugal, Spain, South Korea (Crow & Thomas 1983), Hong Kong (Cody *et al.* 1980, Crow & Thomas 1983), and Singapore and Taiwan (Cody *et al.* 1980). Using only measures of manufacturing, one source seems to suggest that Jordan, Lebanon, Argentina, Chile, the People's Republic of China, India, and North Korea also have attained characteristics of semi-industrialized countries (Cody *et al.* 1980) and could thus be construed as belonging to such an NIC list. It is such uncertainty over inclusion or exclusion of specific countries on such lists that prompts an effort to derive a more systematic procedure for identifying NICs.

In Table 7.1 are listed the primary characteristics of NICs as derived from the literature summarized earlier. In the adjacent column are the variables derived from the 1985 *World development report* (hereafter referred to as the WDR) which most closely approximates each NIC characteristic. These are termed "operational variables" since they are the functional definition of the concept they represent. Data were collected to represent each of the operational variables listed in Table 7.1. While there are difficulties with specific aspects of the WDR data, and while the data for some countries are incomplete (nonmarket economies such as East Europe) or unavailable (Taiwan), the WDR remains one of the most comprehensive and useful sources of international economic data.

Two research assumptions are fundamental to the two-step identification procedure outlined in this essay:

(a) NICs are middle-income states. While one nation, Spain, identified as an NIC by several sources, is categorized by the *World Development Report* as an industrial market economy and is not considered middle-income, there seems to be widespread agreement that NICs are middle-income. Indeed, NICs should have attained levels of per capita GNP *above* the world average. Operationally, the WDR divides the middle-income states into two categories, lower- and upper-middle-income, NICs should have attained an income level above the average for lower-middle-income states.

(b) NICs are characterized by rapid economic and industrial growth and should demonstrate rates of overall growth and, more specifically, industrial and manufacturing growth that are higher than those of most middle-income states. NICs should demonstrate rates of growth at least above the averages for lower-middle-income states.

While these assumptions are based on characteristics of NICs derived from existing literature, there are admittedly debatable issues addressed within each. These assumptions suggest that an NIC must have attained certain levels of wealth and industrial activity that set them apart from the average nation. It is possible that such "above average" thresholds overlook nations that have begun their economic development process but that have not attained middle-income status. However, by employing averages for lower-middle-income countries, the threshold appears low enough to avoid errors of exclusion.

In Step 1 of the identification process, the WDR was used to derive the per capita GNP of all countries. Based on the assumption that NICs are middle-income, all countries identified by the WDR as lower-income (having 1983 per capita GNP of US $430 or less) were eliminated from consideration as potential NICs. No countries with a per capita GNP of US $430 or less would have qualified using the procedures outlined in Step 1. None ranked sufficiently high on the two qualifying variables. For those countries

Table 7.2 Qualifying by level and share of industrial activity.

Country	Per capita GNP (1981)	Percentage of labor force in industry (1981) (%)	Distribution of GDP – percentage in manufacturing (1983) (%)
Averages for lower-middle-income nations	440	17	16
Bolivia	510	24	16
Honduras	670	20	15
Egypt	700	30	27
El Salvador	710	22	15
Morocco	760	21	17
Philippines	760	17	25
Costa Rica	1,020	23	20
Peru	1,040	19	26
Jamaica	1,300	18	19
Dominican Republic	1,370	18	18
Paraguay	1,410	19	16
Ecuador	1,420	17	18
Colombia	1,430	21	17
Taiwan	1,400	37	48
Jordan	1,640	20	15
Malaysia	1,860	16	19
Chile	1,870	19	20
Brazil	1,880	24	27
Republic of Korea	2,010	29	27
Argentina	2,070	28	28
Portugal	2,230	35	35
Mexico	2,240	26	22
South Africa	2,490	29	27
Uruguay	2,490	32	26
Yugoslavia	2,570	35	32
Venezuela	3,840	27	17
Greece	3,920	28	18
Israel	5,370	36	24
Hong Kong	6,000	57	22
Singapore	6,620	39	24
Spain	4,780	40	22
Ireland	5,000	37	29
Italy	6,400	45	29
New Zealand	7,730	35	23
Belgium	9,130	41	25
United Kingdom	9,200	42	18
Austria	9,250	37	27
Netherlands	9,890	45	24
Japan	10,120	39	30
France	10,500	39	25
Finland	10,740	35	23
German FR	11,430	46	36
Australia	11,490	33	20
Denmark	11,570	35	16
Canada	12,310	29	16
Sweden	12,470	34	22
United States	14,110	32	21

Table 7.3 Source, location and description of the variables used in the NIC identification procedure.

Table	Variable
Step 1	
Table 1	Income: per capita GDP, 1983
Table 21	Labour force: percentage of the labor force employed in industry, 1965–81
Table 3	Distribution of GDP: percentage devoted to manufacturing, 1983
Step 2	
Table 21	Labour force: percentage change in the percentage of the labor force employed in industry, 1965–81
Table 3	Structure of production: distribution of GDP: percentage change in the percentage of the total GDP that is in manufacturing, 1965–83
Table 2	Growth of production: average annual growth rate in the percentage of manufacturing as a percentage of the total GDP, 1973–83
Table 1	Reduction in per capita GNP gap: percentage change in difference in per capita GNP between the US and each country, 1974–83
Table 10	Structure of merchandise exports: percentage change in the percentage share of exports which are manufactures, 1965–82

Source: *World Development Report* 1985.

identified by WDR as middle-income or above (a 1983 per capita GNP of US $440 or above), data on the percentage of the labor force employed in industry in 1981 and data depicting the percentage of the total GDP devoted to manufacturing in 1983 were collected. These countries for which the level of industrial employment and manufacturing share of the total GDP were above the average for lower-middle-income states identified in the WDR were selected for further analysis. According to the assumptions established above, these countries "qualified" since they have sufficient levels of industrial employment and manufacturing activity. To qualify a country had to have both a level of industrial employment in excess of 16 percent and at least 15 percent of its total GDP devoted to manufacturing. A list of these countries and the data for per capita GNP, percentage of labor force in industry and percentage of total GDP in manufacturing is provided in Table 7.2.

In Step 2 of the identification process, the data depicting each of the remaining operational variables outlined in Table 7.1 were collected for each of the nations that "qualified" in Step 1. Each of these variables is intended to depict one component of growth or development activity which characterizes an NIC. Three of the variables depict change in the manufacturing sector of the economy. Two of these focus on production, particularly on manufacturing production as a proportion of GDP. Of these two, one measures the average annual growth of manufacturing production between 1973 and 1983; the other focuses on change in the structure of manufacturing production from 1965 and 1983. The third manufacturing variable traces the

Table 7.4 Identifying the NICs: qualifying on growth indicators.

Country	Change in labor force 1965–81 (%)	Change in GDP in manufacturing 1965–81 (%)	Growth rate in manufacturing as percentage of GDP 1973–83 (%)	Change in GNP gap in relation to the US, 1974–83 (%)	Change in manufacturing as percentage of merchandise exports (%)
Average lower-middle-income:	4	1	5.4	1.2	11
Bolivia	4	0	1.7	0	–
Egypt	15	7	9.3	1	−13
Morocco	6	1	4	−1	30
Philippines	1	−2	5	0	45
Costa Rica	3	6	6	−5	27
Peru	0	6	0.4	−4	13
Jamaica	−7	2	−3.6	−8	29
Dominican Republic	5	4	4.4	1	15
Paraguay	0	0	7.4	3	8
Ecuador	−4	0	8.9	3	1
Colombia	1	−1	1.9	2	17
Taiwan	26	16	13.2	6.5	39
Malaysia*	3	9	10.6	3	17
Chile	−2	−4	0.5	1	4
Brazil	7	1	4.2	−1	31
Republic of Korea	16	9	11.8	7	32
Argentina	−6	−5	−1.8	−14	18
Portugal	4	6	4.5	−7	13
Mexico	5	1	5.5	1	−4
South Africa	−1	3	–	0	42
Uruguay	2	5	3.4	2	1
Yugoslavia	14	−4	8.2	−1	21
Venezuela	3	–	3.7	0	−3
Greece	6	2	2.7	−2	37
Israel	1	15	6.1	−13	15
Hong Kong	3	−2	6.9	20	5
Singapore	13	9	7.9	15	22
Spain	5	0.6	4.1	4	31
Ireland	9	0	–	−1	31
Italy	3	−2	3.7	3	6
New Zealand	−1	–	–	−7	19
Belgium	−5	−5	1	−13	−2
United Kingdom	−4	−12	−1.9	14	−16
Austria	−8	−6	2.7	5	10
Netherlands	2	−10	1.9	−3	−6
Japan	7	−2	6.6	14	6
France	−1	−4	2.8	−4	3
Finland	2	2	3.6	14	20
Germany, FR	−2	−5	1.8	−8	−1
Australia	−6	−8	1.5	9	8
Denmark	−2	−4	2.3	−6	13
Canada	−4	−4	0.8	−5	17
Sweden	−9	−6	−0.1	−13	11
United States	−4	−8	1.4	0	5

* Although Malaysia failed to qualify as an NIC in Step 1 (see text) it is included here because the growth indicators suggest that Malaysia may soon qualify as an NIC.

change between 1965 and 1982 in the share of merchandise exports devoted to manufactured products.

The fourth and fifth variables trace other aspects of economic change. One traces the change in the structure of the labor force by measuring the growth of employment in the industrial sector between 1965 and 1981. Finally, the fifth variable is selected to measure how well the economic growth in each of these countries is raising the overall level of income relative to the US. In this instance the US is taken as a surrogate of the composite of wealthy industrial countries. Table 7.3 provides a broader definition of each of the variables used in this study.

Any nation that is exceptionally active in each of the areas of economic activity depicted by these variables certainly demonstrates the levels of growth and development that characterize an NIC. We employ the standard that an NIC must be at least above the average of lower-middle-income states identified by the WDR on each of these operational variables. In this instance there are five variables on which each country must qualify. NICs should demonstrate sufficiently strong patterns of growth in industry, manufacturing, and wealth to rank above average for lower-middle-income countries on all five criteria.

Table 7.4 lists all the countries that qualified according to the procedures outlined in Step 1. The data representing each of the five operational variables used in Step 2 are provided for each country. The average values for countries listed by WDR as lower-middle-income are provided. Table 7.5 lists the countries that rank above average on all five variables and those that rank above average on four out of the five operational variables.

The conditional NICs

The countries identified by this classification system might just as appropriately be called "newly manufacturing market economies." They are the élite among the world's middle-income countries. Because manufacturing is typically the most dynamic part of the industrial sector and because economic growth is often difficult to achieve and maintain, very few countries successfully qualify on all diagnostic criteria. The three nations that qualify unconditionally are those countries with the most dynamic, progressive, and promising young industrial economies. Many of the characteristics that best identify NICs are illustrated by the three rapidly growing economies of Taiwan, the Republic of Korea, and Singapore.

Many nations, including the high-income oil exporters and the industrial market economies, meet several of the criteria. The oil-exporting countries had a high percentage of the labor force in industry and had greatly reduced the per capita GNP gap between themselves and the United States during the 1974–83 period. Most of their employment in industry results from mining

Table 7.5 The final list of NICs identified.

Nations that meet all five criteria:
Taiwan
Singapore
Korea, Republic of
Nations that meet four of the five criteria:
Malaysia
Egypt
Dominican Republic
Jordan
Portugal
Mexico
Spain

and the improvements in the GNP gap result from rapidly rising oil revenues. The contribution of manufacturing to the GNP, however, has been only about one-third of that in the NICs.

Most of the industrial market economies showed very slow if any growth in the manufacturing sector. As the economies of the industrial nations continue to evolve, service-sector employment typically increases at the expense of manufacturing employment. In the United States, for example, employment in industry decreased by 4% (Table 7.4). Even those industrial market economies with the greatest increases in the industrial labor force, e.g. Ireland (9%), Japan (7%), and Spain (5%), look anemic compared with Taiwan (26%), the Republic of Korea (16%), or Singapore (13%).

Most of the World Bank's 39 lower-middle-income countries have very slowly growing or stagnant economies. Egypt and the Dominican Republic are notable exceptions. Egyptian manufacturing and industrial growth are almost sufficient to classify Egypt as an NIC: however, the structure of merchandise exports shows that the importance of manufacturing among the export goods and commodities is decreasing. Egyptian exports of limited oil resources claimed nearly 70% of all merchandise exports by 1982. The relative importance of manufacturing has declined by 13%. With a per capita GNP of only $700 in 1983 Egypt is certainly a poor country. Some of Egypt's growth as reflected in this classification results from adding modest increases to a small baseline.

The Dominican Republic has twice the GNP per capita ($1370) of Egypt, but it also has a limited resource base. Agricultural products, principally coffee and sugar, dominate merchandise exports. The Dominican Republic does not yet have enough growth in manufacturing to qualify as an NIC but like Egypt it appears to be a potential candidate.

Although frequently assumed to be an NIC, Hong Kong no longer seems to qualify. Three indicators of industrial growth – increase in the percentage of labor force in industry, increase in the share of the GDP that is manufacturing, and increase in the share of manufacturing exports – suggest that the industrial growth is too slow for Hong Kong to be an NIC. For example, in 1985 Hong Kong's economy grew by only 0.8%. Hong Kong's industrial growth has slowed because it is nearly fully industrialized. Fifty-seven percent of the work force is employed in industry and those working in manufacturing produce 92% of Hong Kong's exports. It appears that Hong Kong is not a newly industrializing country but an already quite industrialized city-state.

South Africa fails to qualify on at least two criteria. The percentage of the labor force employed in industry actually *declined* during the 1965–81 period, and the per capita GNP gap with the United States remained as wide in 1983 as it had been in 1974. In 1985 under the burden of record inflation (20.7%), South Africa's real GNP declined by 1%. Lack of or sluggish improvement in each of these two characteristics can be attributed to a feudalistic social system that has denied equitable educational, employment, and investment opportunities to the black population. With black employment at 30% of the black labor force moderate economic expansion with increases in black employment is of greater immediate importance than curbing inflation. Consequently, living standards in South Africa are likely to fall over the next few years.

Brazil's economy qualified on all but two criteria. Struggling under a $104 billion foreign debt and annual inflation of 220%, the per capita GNP gap with the United States has actually widened. Although systematic comparisons with other countries since 1983 were not attempted in this classification, Brazil did accomplish sizeable growth during the 1983–5 period. The GDP increased by 4.4% in 1984. Exports were up 23% overall. Primary exports (coffee, soya, iron, and sugar) rose only 14%, while manufactured exports rose by 32%. Despite its oil imports Brazil recorded a $13.1 billion trade surplus in 1984 and the expectations for 1985 appear equally promising. Real growth in Brazil's GDP is expected to equal 5% in 1985 and to help to recover the paltry GDP growth (or decline) during the 1981–3 period. If the present boom can be sustained, Brazil could qualify as an NIC by 1988.

Mexico almost qualifies but manufactured products actually declined as a percentage of total exports between 1965 and 1982. That trend has continued through 1985. Today 60% of Mexico's export earnings result from the sale of oil. With a $96 billion foreign debt, an annual inflation rate close to 60% for 1984 and 1985, 900,000 young workers entering the job market in 1985, a catastrophic earthquake, erratic swings in the world price of oil, and a vanishing trade surplus, it seems likely that Mexico's economy can do little more than stagnate. The greatest prospects for the Mexican economy lie in its enviable location adjacent to the world's largest market for imported manufactured products, and the considerable wage advantage it enjoys compared to its northern neighbors. Mexico's need for foreign technology and investment has suffered from an economic oligopoly and a political system that has been practically xenophobic. Foreign investment in Mexico is less than 5% of all investment (compared with 40% in Canada).

In some respects the Portuguese situation is analogous to Mexico's. The European Common Market holds out considerable possibilities for expanding the sales of Portuguese manufactured goods. Portugal's comparative advantage is now in labor-intensive industries such as textiles, clothing, and shoes but possibilities for expansion include agricultural and textile machinery, heavy machine tools, and castings.

Jordan, with a population of only 3.2 million people and the least wealthy of the upper-middle-income countries, has easily shown growth during the last 15 years. Jordan fails to qualify as an NIC because the contribution of manufacturing to the GDP has decreased during the 1965–83 period. Much of Jordan's recent economic growth appears to be occurring in the tertiary sector. Although 50% of Jordan's merchandise exports in 1983 were manufactured goods, total exports in 1983 were a relatively small $739 million (compared with Israel's $5.1 billion). Also, Jordan has increased its external public debt from 23.5% of the GNP in 1970 to 47.9% in 1983. The debt service is now over 5% of the GNP. In 1983 Jordan had a trade deficit of $2.5 billion which is likely to make continued improvement difficult.

The percentage of Malaysia's labor force employed in industry has been less than the average for lower-middle-income nations and consequently Malaysia does not currently qualify as an NIC. All of Malaysia's growth indicators, however, suggest a robust increase in the level of industrialization and inclusion of Malaysia among the NICs during the next few years seems likely.

Malaysia is an export-dominated, resource-rich democracy that has benefited from an annual GNP growth of 7.9% during the period 1980–4. The manufacturing sector is largely financed with foreign capital, in consequence of which Malaysia is often described as "neocolonial." In 1984 the leading exports were: oil and gas, 29%; tin, rubber, and other raw materials, 22%; machinery and transportation equipment, 18%; and animal and vegetable oil and fats, 15%. The dominance of tin and rubber in the export economy has declined sharply from their 90% share in 1950. The nascent manufacturing sector, especially chemicals, petroleum, metals, machinery, and transportation equipment, has received relatively little protection, has been forced to be efficient, and has acquired considerable export potential. Malaysian manufactured exports are generally labor-intensive products for which Malaysia has a comparative advantage.

The prospects for Malaysian economic growth appear quite promising, especially in the short term; however, several problems present some threat to continued rapid growth. Commodities such as oil, tin, and rubber are no longer expected to be the bonanza they once were. Direct Malaysian trade with the large and growing Chinese market has not been increasing despite attempts to bypass middlemen in Hong Kong and Singapore. While China's share of Malaysia's trade has stagnated, Singapore's has increased. Also, the income generated by Malaysian industrial growth over the last 15 years has been very inequitably distributed. Urban concentrations of ethnic Chinese dominate employment in administration and managerial work, sales, and manufacturing, while the rural Malay population is trapped in primary-sector employment and benefits little from the industrializing economy.

The NICS

Taiwan, the Republic of Korea, and Singapore meet all the criteria for inclusion as NICS. Taiwan had the world's fastest-growing economy in 1984, notching up a GNP growth of 11%. Singapore and South Korea grew by an average of 8%. The principal assets common to these young dragons have been discipline, reliability, and political and economic stability. The keystones of their success have been aggressive international trade, rising education and labor skills, a growing number of middle-class consumers, and a receptive investment climate. During 1985 these economies suffered a mild slowdown in their rates of growth because of slow or stagnating exports. Manufactured exports from these four countries are purchased mostly by the United States. For example, the US market accounted for all of Taiwan's export growth and 80% of South Korea's during 1984. Currencies in the NICS have been effectively tied to the US dollar and subsequently their exports on other markets have become less competitive as the strength of the dollar rose. During the first half of 1985, rising protectionist sentiment and slower growth in the US economy reduced export growth and consequently GNP growth among the NICS.

Taiwan's economy faces a critical period in its development. Taiwan's growth has been based on an increasingly favorable trade balance which reached an $8.5 billion surplus in 1984. Most of Taiwan's exports are labor-intensive, low-technology products such as cloth and cotton clothes. United States sales of Taiwanese textiles and clothing were $2.3 billion in 1984. Ironically, wages in Taiwan have risen during the last decade to the extent that Taiwan now faces low-wage competition in textiles from the People's Republic of China, India, and Indonesia. Even dragons get their tails chased. To counter this trend Taiwan is attempting to follow the Japanese model and leapfrog these interlopers by switching to high-technology capital-intensive manufacturing. Unlike Singapore or even South Korea, Taiwan is finding it politically difficult to compensate by expanding indirect trade of low-priced electronics to China (via Hong Kong). In Taipei there is fear that China may deliberately entrap Taiwan through overdependence. Taiwanese trade with China is expanding rapidly nevertheless and approached one billion (US dollars) in 1985.

With the exception of Taiwan, South Korea recorded the greatest increase in the percentage of the labor force employed in industry, and the greatest growth rate for manufacturing of any country in the study (Table 7.4). Although the trend over the last 15 years clearly identifies South Korea as an NIC, the South Korean economy has shown some minor instability. Poor grain harvests, minor political unrest, and 35% per year inflation sent the economy crashing (real GNP fell by 5%) in 1980. By 1985 inflation had lowered to 2% per year by restricting the money supply and by reducing government debt by increasing taxes. The result has been a bustling economic expansion of 5–10% per year in the GNP. South Korean exports are closely tied to the US economy. Thirty-five percent of South Korean exports went to the United States in 1984.

Three types of relatively low-technology goods comprise almost half of South Korean exports: shipbuilding, shoes, and textiles. Recent heavy investment in semiconductors and automobiles signals an attempt to shift up-market into more lucrative high-technology products such as cars and electronics. Two of South Korea's largest conglomerates, Daewoo and Hyunda, are marketing cars in the United States. Samsung has already established a growing reputation for electronic consumer goods in the American market. Prospects for the Republic of Korea appear bright, especially through the current cycle of American economic growth. Like Taiwanese companies these firms not only seek to produce sophisticated products but also to be innovative originators of improved technology.

Singapore is a 238-square-mile republic that is heavily urbanized, clean, efficient, and most fortuitously situated along one of the world's busiest shipping routes, the Strait of Malacca. Under the leadership of Lee Kuan Yew since 1958, Singapore has increasingly sought and developed an export economy. Today Singapore exports chemicals, oil-drilling rigs, petroleum products (although Singapore itself has no oil reserves), batteries, pharmaceuticals, and engines. The American market takes 20% of Singapore's exports.

Over the last 20 years, since independence from Britain, the standard of living has risen faster in Singapore than anywhere else on earth. For the past 15 years the Singaporean GNP has fluctuated between growth rates of 5% and 14%. The world oil glut has begun to dampen demand for rig building, oil refining, and ship repair, so that Singapore experienced an economic slowdown during the last two years.

Lee Kuan Yew's heavy emphasis on manufacturing has been a most prudent and successful strategy over the past two decades. Singapore, however, is now facing fierce competition from neighboring Asian nations. For example, Malaysia and Indonesia in building their own refineries have left much of Singapore's refining capacity redundant. Unlike Taiwan and

South Korea, Singapore has not chosen to support research, development and innovation in industry. With only one university for five million people, Singapore offers higher education to only 8 or 9% of the population. The stultifying effects of too few scientists and engineers leave Singaporean manufacturing relegated to slug it out with low-wage competition from China, Malaysia, and Indonesia.

Conclusions

The steps employed in this classification provide a yardstick by which the economic progress in various nations can be systematically compared. The system is designed to be an accurate, orderly, and rational introduction to the study of newly industrializing countries. The indicators of development that are used in the classification procedure are easily obtained from the World Bank each year. Although several of the criteria used here are trends averaged over the previous decade or two, the economies of several Third World countries are sufficiently dynamic to make annual revisions in the list to NICs worthwhile.

The newly industrializing countries have followed widely differing patterns of development. While diverse fiscal and labor policies, foreign investment policies, trade barriers, and export incentives have led to differences in industrial structures, the progress of these nations in developing a manufacturing export economy has been nothing less than miraculous over the last 20 years. East Asian countries such as Taiwan, South Korea, Singapore, and Malaysia have been particularly successful. As their economies continue to grow, these countries will play an increasingly prominent rôle in American foreign trade and in the economic development of Asia. To date their growth has been based on their most ample resource, labor. Their continued economic success will depend upon their ability to maintain political stability, to avoid large military expenditures, to balance industrial development with agriculture, mining, and forestry, and to maintain an internationally competitive export economy. Taiwan and the Republic of Korea in particular are poised on the threshold of high-technology capital-intensive manufacturing. Success in making this transition will help to insure continued growth and increasing prosperity among the current and future newly industrializing countries.

References

Cody, J., H. Hughes, & D. Wall 1980. *Policies for industrial progress in developing countries*. New York: Oxford University Press.

Crow, B. & A. Thomas 1983. *Third World atlas*. Milton Keynes: Open University Press.

Dadzie, K. K. S. 1980. Economic development. In *Economic development*. San Francisco: W. H. Freeman.
Forbes, D. K. 1984. *The geography of underdevelopment*. Baltimore: Johns Hopkins University Press.
Ingalls, G. L. & W. E. Martin 1983. *Exploring world regions*. Dubuque, Iowa: Kendall Hunt.
Jumper, S. R., T. L. Bell, & B. A. Ralston 1980. *Economic growth and disparities: a world view*. Englewood Cliffs: Prentice Hall.
Kuznets, S. 1965. *Economic growth and structure*. New York: Norton.
Population Reference Bureau 1976. *World population data sheet*. Washington, DC: Population Reference Bureau.
Reynolds, L. G. 1985. *Economic growth in the Third World, 1850–1980*. New Haven, Conn.: Yale University Press.
Rostow, W. W. 1978. *The world economy: history and prospect*. Austin, Texas: University of Texas Press.
OECD 1981. *The impact of the newly industrializing countries*. Paris: Organization for Economic Cooperation and Development.
The World Bank 1985. *World development report*. New York: Oxford University Press.
The World Bank 1984. *World development report*. New York: Oxford University Press.
The World Bank 1981. *World development report*. New York: Oxford University Press.
The World Bank 1980. *World development report*. New York: Oxford University Press.
The World Bank 1978. *World development report*. Washington, DC: The World Bank.
The World Bank 1980. *World tables*, 2nd edn. Washington, DC: The World Bank.

8 *The poorest of us all*

LAURENCE GRAMBOW WOLF

In western Europe and the United States unsold foodstocks accumulate. Modern technology and interest-group politics have produced an embarrassment of riches. As this is being written, severe famine ravages many millions of luckless people, especially in the eastern reaches of the vast semi-arid region that lies between the Sahara and the tropical forests of Africa. The unequal distribution of the world's goods has rarely been more starkly revealed.

Unequal distribution, as such, comes as no surprise to a geographer. It is a basic fact of life and provides the *raison d'être* for our entire profession. Others may rely heavily upon the construct "all other things being equal," upon "ubiquities," and upon scarcities which rarely seem to occur in any particular places. Nothing at all is ubiquitous in geography: scarcities are space-specific as well as income-related, and all things are unequal in their occurrence over space as well as time. Thereby hangs the differentiation of places, regions, and countries, and therefore of people, one from another. Hence the need to pay attention to the location of one's objects of study and to seek systemic explanations of the significant differences and similarities that are found from place to place.

This essay focuses upon the Third World, a term which is here taken to mean all the world outside Europe, Anglo-America, Anglo-Oceania, Japan and the Soviet Union, however else it may be defined by others. (Anglo-America is the United States and Canada; Anglo-Oceania consists of Australia and New Zealand.) Although South Africa is sometimes excluded, the historical, cultural, and economic conditions of the overwhelming majority of its population warrants inclusion. Israel is included on the basis of its location and the origin of the majority of its people; while Portugal and the Balkans are excluded on the same basis, though in terms of purely economic criteria, such as their gross national product (GNP), they could qualify for inclusion. Singapore, Hong Kong, and Macao are omitted. These city-states are not countries in any normal sense. Their experiences are therefore not of great consequence for Third World countries. The term "core" as used herein refers specifically to western and northern Europe, Japan, and Anglo-America.

Three-quarters of the world's people now live in the Third World, a slight increase over 72% a decade ago. Is the wellbeing of this population improving? Is the gap between the Third World and the core decreasing? If the modern industrial economy is diffusing successfully, the answer to both questions should be "yes." The Third World includes so much of the world, however, a geographer would be much surprised if the same answers applied uniformly. What regionality characterizes the development of the Third World over the past decade?

The modern industrial economy operates on several assumptions, one of which is that the entire world can be developed. "Development" can be taken to mean that somehow, by some supposedly normal process of diffusion, the same high level of industrial productivity and its consequent amenities that prevail in the core will encompass all the world's lands and peoples in due time. In a more limited and practical sense, it means that subsistence economies and local market economies, when penetrated by the global market economy, will progress – will be developed – regardless of the level of sophistication the native economy may have had. Anything in its natural or previous state is deprecated. Any commercialization that reduces traditional autonomy and traditional networks is seen as progress, even if it is only the exploitation (perhaps temporary) of mineral deposits, the creation of plantations of bananas or copra, or the building of tourist resorts.

Heavily dependent in its early stages on brute force, as the history of the Western European empires attests, the modern market economy now rests on an abundance of material lures. These are quite sufficient and efficient in orienting the educated élites of Third World countries toward the promotion of policies of economic development inspired by, if not actually created by, the core. Whatever criticisms can be made of the modern industrial world economy – and there are many that are legitimate and profound – it has produced an unprecedented degree of material wellbeing for a majority of the populace of the core countries. The internal compulsion of this economic system to expand is now aided by the desire to participate in it on the part of those who have, or hope for, an opportunity to prosper thereby. Western empires were not swept away by a great mass revulsion on the part of the populace of the colonies. On the contrary, a moment of weakness was seized upon. Colonial overlords had exhausted themselves in a worldwide war. Indigenous nationalist anti-colonial change-agents took over precisely because they were of the opinion that benefits of the modern Western system of production could be gained more rapidly under native rather than foreign leadership. The "revolution of rising expectations" that followed World War II is still with us. It is precisely the most nationalist change-agents in the Third World who have become the most effective destroyers of their own societies' traditional ways of living. Not until the rise of the ayatollahs did an anti-modern backlash become politically significant, and it remains to be seen how this will work itself out.

One of the commonly used measures of national economies is the gross national product. Because countries with similar GNPs vary greatly in the size of their populations, annual per capita GNP should be used so that total national economic production is related directly to national population. There are serious flaws in this measure, but there is *no* universally accepted or completely flawless measure, and GNP data are widely available for a long list of countries. The GNP is taken herein more as an indication of the penetration of Third World national economies by the modern monetary-exchange economy than as a measure of economic production. The less developed a country, the more it is likely that the GNP indicates a level of activity involving only a portion of the populace. In the more extreme cases, even a majority of the people still derive their livelihood entirely or mostly from subsistence or local market economies. Therefore, the GNP, like any measure of the monetary-exchange market economy, understates the level of economic activity in underdeveloped countries.

Third World countries have been classified into six categories pegged to the world annual per capita GNP figure: Category I, up to one-fourth of the world per capita GNP value; Category II, one-fourth to one-half; Category III, one-half up to the world value; Category IV, from the world value to twice that; Category V, from twice to four times the world value; and Category VI, four times the world value or more. The world annual per capita GNP was $940 in the early 1970s and $2,760 in the early 1980s (Table 8.1). This categorization provides a degree of generalization between the extremes of dealing with each country separately or with all of them in the aggregate. (These data are derived from the 1975 and 1985 *World population data sheet* of the Population Reference Bureau.)

Improvement in the economic position of the Third World, if integration into the world economy can be taken as improvement, can be indicated by an upward redistribution of national economies among these categories. The lower ones (I and II) should lose and the higher ones (V and VI) should gain countries. There has been little relative improvement, as the Table 8.1 indicates. More than half the world's population was in Category I in the early 1970s and is still there in the early 1980s. The middle categories (III and IV) which now have 11% of the world's population had 8% ten or so years earlier. The high categories (V and VI) have doubled their population, but it remains a very small fraction of the whole nevertheless. The Third World grew during the decade by almost 800 million people. However, 55% of that increase occurred in Category I alone, and 26% in Category III, while Category IV declined by 41% from the mere 43 million it had at the start of the decade. Individual countries have improved their status, but, overall, improvement in the decade under consideration is not great.

Countries do move from category to category, and 35 did so: 20 for the better and 15 otherwise. Nicaragua, for instance, increased its per capita GNP from $470 to $900, but this was insufficient to keep pace with the world economy and it went from Category III to Category II, as did three others, while eight went from II to I. The greatest advances came in a few Middle

Table 8.1 Number of countries, population, and area by GNP category.

Annual per capita GNP category*	Number of countries		Population (millions)		World population (%)		World land area (%)	
	1980s	1970s	1980s	1970s	1980s	1970s	1980s	1970s
I	52	47	2,633.6	2,197.1	54.4	55.4	27.9	28.0
II	24	33	432.7	288.9	8.9	7.3	7.8	10.3
Low	76	80	3,066.3	2,486.0	63.3	62.7	35.7	38.3
III	20	20	491.3	280.8	10.1	7.1	17.1	14.5
IV	9	7	24.9	42.5	0.5	1.1	1.0	4.1
Middle	29	27	516.2	323.3	10.6	8.2	18.1	18.6
V	4	5	6.8	6.8	0.1	0.2	1.5	0.1
VI	4	1	14.7	1.1	0.3	n†	1.7	n†
High	8	6	21.5	7.9	0.4	0.2	3.2	0.1
no data	40	40	55.7	47.0	1.2	1.2	0.6	0.6
Total	153	153	3,659.7	2,864.2	75.5	72.3	57.6	57.6

* Dollar value of GNP categories:

1980s				1970s			
I	110–689	IV	2,760–5,519	I	60–234	IV	940–1,879
II	690–1.579	V	5,520–11,039	II	235–469	V	1,879–3,759
III	1,380–2,759	VI	11,040–21,340	III	470–939	VI	3,760–4,090

† n: negligible (less than 0.1%).

Eastern petroleum-producing countries (Saudi Arabia, Bahrain, and Oman had increases of more than 1,000%).

The experience of such rapid, drastic increase is rather irrelevant for the rest of the Third World countries, because they do not have any resources comparable in importance to petroleum which could be exploited to bring in such wealth so rapidly. Nigeria, even with its petroleum, has a large and growing population and therefore managed to advance from a per capita GNP of $130 (Category I) to $760 (Category II) – an improvement, to be sure, but one that hardly matches that of the sparsely populated Arab states. An attempt at a bauxite cartel failed some years ago and the prospects for cacao, fiber, or other cartels among Third World raw-material-producing countries seem dim indeed.

Among the more advanced Third World countries, Argentina (going from $1,290 to $2,030) has slipped from Category IV to III. Its business class has often shown remarkable political ineffectiveness in the face of working-class dissatisfaction and confusion and the arrogance and ineptitude of the military. One could cite many other examples. The problems of the Third World are in part environmental, they are very much economic, and they are inescapably political. On the positive side, 13 widely-scattered non-

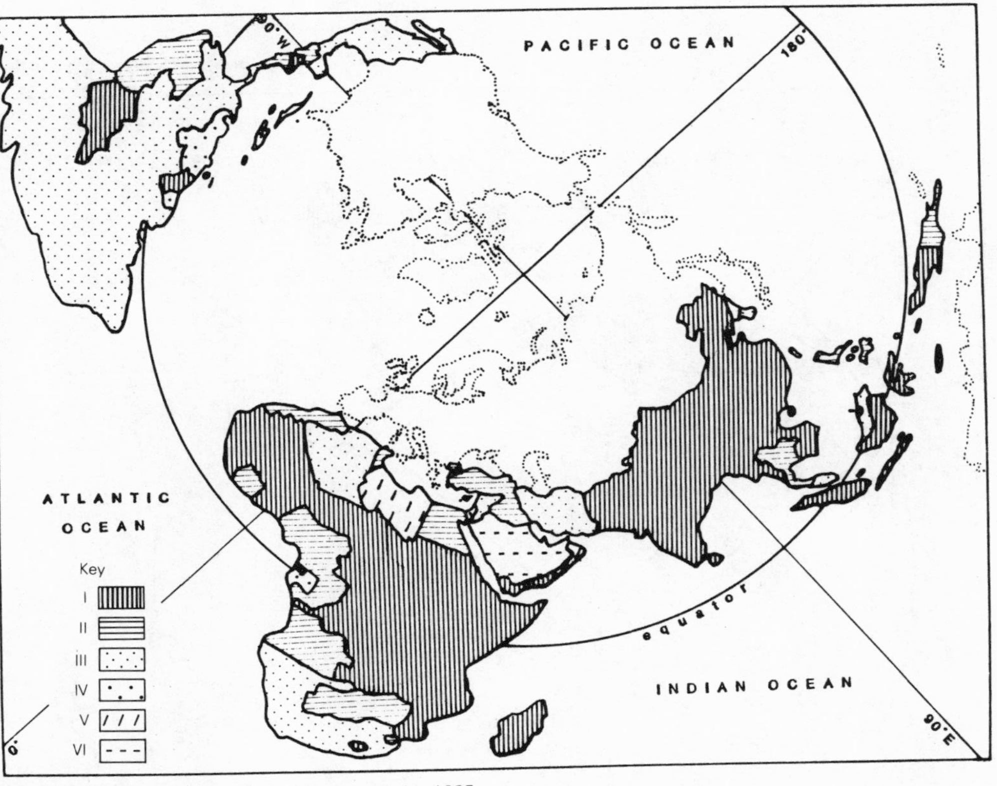

Figure 8.1 Annual per capita GNP categories 1985.

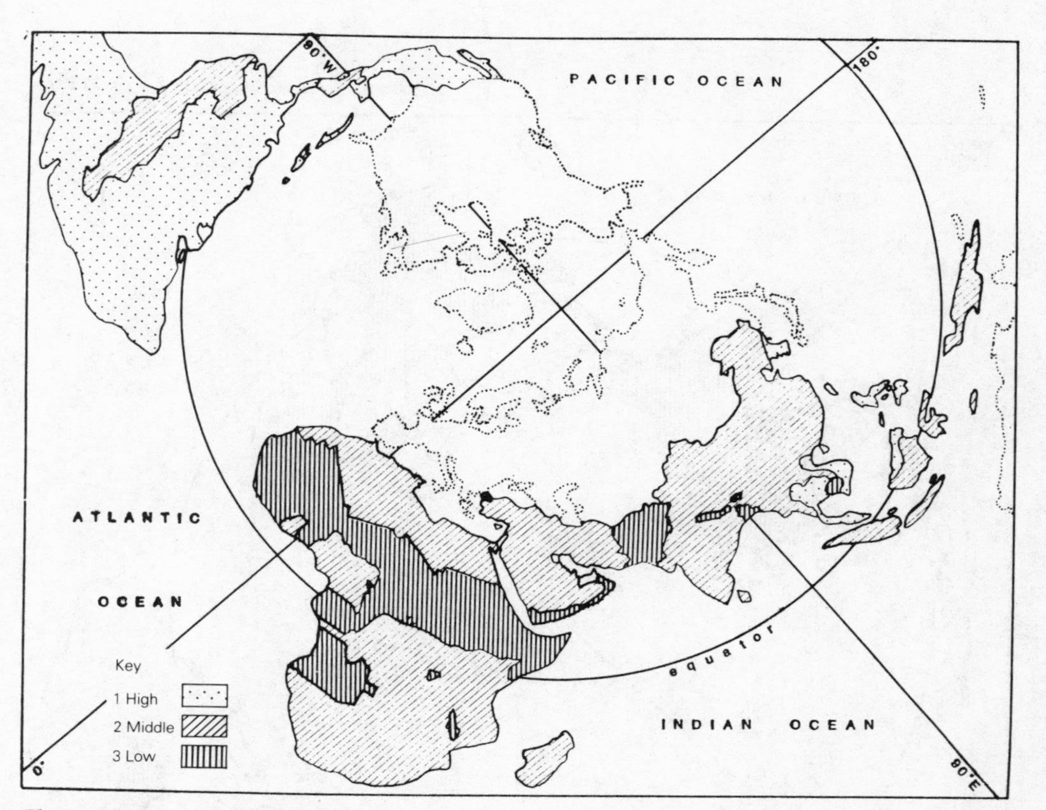

Figure 8.2 PQLI Categories 1985.

petroleum-producing countries had per capita GNP increases in excess of 300%, as compared to the world rate of +194%. However, they contain only 2% of the world's population. Their rather random location seems to indicate that they are special cases.

Really low values (Category I), as a comparison of the maps in Figures 8.1 & 8.2 reveals, are very much a phenomenon of tropical Africa and of eastern, southern and southeastern Asia, with Category II countries tending to be located peripheral to these. Latin America, however, is very much a region of Category III. Only Bolivia and Haiti are consistently in Category I. This is a striking confirmation of the fact that the Third World contains major regional differences. Latin America lacks the huge population concentrations of southern and eastern Asia, where relatively small increases in population can consume relatively large increases in production. This region, also, was generally penetrated and developed (peripheralized) sooner than Saharan or transSaharan Africa, some of which was still being "discovered" and subdued at the beginning of this century.

One of the ways of getting beyond the GNP for a measure closer to human wellbeing is the Physical Quality-of-Life Index (PQLI) devised by Morris D. Morris (Morris 1979). This has three components: infant mortality, life expectancy at age one, and literacy. Each of these is discussed briefly before the combined PQLI scores are examined.

In core countries, parents are shocked when pregnancy does not result in a viable infant. There are many Third World countries in which such success is normally quite problematical. Afghanistan is currently the worst case, with an infant mortality ratio of 205 per 1,000. The world rate is 81. Of the 128 Third World countries for which data are available from the early 1980s, 67 had infant mortality rates higher than the world figure. Of these, only five were in Latin America. Eleven had experienced an increase in infant mortality, and four actually had more than twice the world figure (Afghanistan, Sierra Leone, Gambia, and Malawi). Twenty-eight others, mostly in Africa, had one and one-half times the world figure. In general, there has been improvement in infant mortality rates around the world, but most Third World countries have a long way to go to reach the rates of 6–12 per 1,000 which prevail in many core countries.

Life expectancy at age one gets beyond the problem of mortality among infants. The world figure for the early 1980s was 61. In Sweden life expectancy at age one is almost 76 and the core countries generally have expectancies of 70 or more. The worst-off Third World countries have expected life spans as much as 40% shorter! When the data are mapped, the concentration of short life-span expectancies in Africa is obvious. Of the 21 countries with values of only 41 to 50, all but three are African. Of the 42 in the 51–60 range, ten are in Asia, two in Latin America, and 27 in Africa. Life expectancies of 61 and over occur (with two exceptions) throughout Latin America, in North Africa, in the Middle East, in China, and in some

Table 8.2 PQLIS and GNP categories.

PQLI	GNP category I	II	III	IV	V	VI	0*	Sum
10–19	3	–	–	–	–	–	–	3
20–29	15	1	–	–	–	–	1	17
30–39	10	2	–	1	1	–	–	14
40–49	7	6	1	–	–	1	1	16
50–59	9	4	3	–	1	–	1	18
60–69	5	5	2	–	1	1	–	14
70–79	1	3	7	2	–	1	–	14
80–89	2	6	7	5	1	3	5	29
90–99	–	3	1	1	1	–	1	7
Totals	52	30	21	9	5	6	9	131

* Countries for which there were PQLI data but no GNP data.

Table 8.3 PQLIS and GNP categories, African countries.

PQLI	GNP category I	II	III	IV	V	VI	0*	Sum
10–19	2	–	–	–	–	–	–	2
20–29	13	1	–	–	–	–	–	14
30–39	5	2	–	1	–	–	–	8
40–49	4	5	1	–	–	–	–	10
50–59	8	3	2	–	1	–	–	14
60–69	–	1	–	–	–	–	–	1
70–79	–	–	–	1	–	–	–	1
80–89	–	1	1	–	–	–	–	2
90–99	–	–	–	–	–	–	–	0
Totals	32	13	4	2	1	–	–	52

* Countries for which there were PQLI data but no GNP data.

Table 8.4 PQLIS and GNP categories, Latin American countries.

PQLI	GNP category I	II	III	IV	V	VI	0*	Sum
10–19	–	–	–	–	–	–	–	0
20–29	–	–	–	–	–	–	–	0
30–39	–	–	–	–	–	–	–	0
40–49	1	–	–	–	–	–	–	1
50–59	1	–	–	–	–	–	–	1
60–69	1	3	1	–	–	–	–	5
70–79	–	1	4	–	–	–	–	5
80–89	1	5	4	4	–	–	2	17
90–99	1	3	1	1	1	–	1	7
Totals	5	12	10	5	1	–	3	36

* Countries for which there were PQLI data but no GNP data.

of the southeast Asian states. Most of the core-level values occur in Latin America.

Literacy is the third component of Morris's PQLI (UN 1982, Kurian 1982). Once again, Latin America stands in contrast to Africa and South Asia. Most African countries have literacy rates from 10 to 60, with the median near 30; while only five Latin American countries have rates that low, and the regional median is in the 80s.

The values for two of the three components are turned into scores from 0 to 100, the highest mortality figure being close to 0, as is the lowest life-expectancy value. The literacy rates need no transformation. These scores are then averaged to produce the PQLI for each country. Giving each component equal weight is a value judgement, but there seems to be no compelling reason to exercise some other judgement. The PQLIS range from 11 (Sierra Leone) to 93 (Cuba, Grenada, Barbados) for the 131 countries for which data were available. The low PQLIS tend to occur in "low income" countries (Table 8.2). Of the 82 "low income" countries, or those with slight integration into the world market economy (Categories I and II), 31 have PQLIS of less than 40 and only 15 are over 70. The contrast of Africa to Latin America remains clear. Of the 52 African countries, only four have indexes over 60, while only two of the 36 Latin American countries have values below that mark (Table 8.3 & 8.4).

The Third World, taken in total, experienced improvement in quality of life as measured with Morris's index. Eighty-four countries had an average increase of 11 index points. The remaining countries suffered a decline which averaged six index points.

Improvements in the PQLI do not seem to be related to GNP or to the earlier PQLI values. However, increases of more than 10 points are more numerous in Africa than elsewhere. Of the 38 countries with scores rising ten or more points, 13 rose from scores of less than 30, indicating that improvement is possible in countries with low PQLIS. Even sizeable improvement is not limited to countries that already have a good base from which to work.

The rate at which population is increasing needs also to be considered. The doubling time for Third World countries at current rates of increase is a mere 34 years. Only 19 of the 130 countries have rates lower than the world rate of 1.7% per year, and more than half the countries have rates that exceed 1.5 times the world rate. Rapidly increasing populations are the norm, astonishingly so in the extreme case of Kenya where the annual rate exceeds 4%, threatening a doubling in 17 years! According to the theory of the demographic transition, the onset of industrialization or modernization should greatly increase the growth rate, but eventually urbanization and higher levels of living should be accompanied by low birth rates and the rate of increase should then subside toward stability. This trend is apparent in the data (Tables 8.5 & 8.6) when per capita GNP categories are related to population growth-rate categories. Growth rates in excess of 2.4 occur in

Table 8.5 Annual population growth rates and GNP categories.

Annual per capita GNP categories	Annual population growth rates, 1984: less than 1.7%	1.7% to 2.4%	2.5% and over	Totals: absolute	Totals: relative
I	3	16	33	52	40%
II	2	4	18	24	18.5
Low	5	20	51	76	58.5
III	5	5	10	20	15.4
IV	3	5	1	9	6.9
Middle	8	10	11	29	22.3
V	1	–	3	4	3.1
VI	–	1	3	4	3.1
High	1	1	6	8	6.2
Subtotals	14	31	68	113	86.9
0*	5	10	2	17	13.1
Totals:					
absolute	19	41	70	130	100.0
relative	15%	32	54	100	–

* Countries for which there are growth-rate data but not GNP data.

64% of the Category I countries and this declines to only 11% in the Category IV countries. Nevertheless, the trend does not hold completely: the high category countries (V and VI) have high growth rates. Possibly their wealth is too recent or is not distributed in such a way as to affect life-styles sufficiently, or some other factor may be preventing the expected changes in demographic trends.

One would expect that as PQLIS increase, population growth rates would decline. Lower infant mortality could conceivably contribute to a decline in fertility that could exceed declines in mortality by reducing the need for many offspring and reducing the growth rate. Greater life spans, however, may not have the same effect. If the social-welfare infrastructure and the condition of the economy do not provide a sense of security as one ages, then the need remains to have many children in the hope that some will be around to assist when the infirmities of old age set in. Whatever the reasons, countries with moderate PQLIS (40 to 69) are overwhelmingly countries with very high population growth rates. Only a quarter of the countries with high PQLIS have such high rates as 2.5 or greater, so here lower rates and higher PQLIS begin to appear, but most of these are tiny countries. The possibility that their experience is a harbinger of that of the rest is indeed problematic.

The prospects for the Third World do not appear especially hopeful. Despite the progress that has been made in most of the countries, including high rates of change in a few countries, most of the population seems not to

Table 8.6 Annual population growth rates and PQLIS.

PQLI	Annual population growth rates, 1984			Totals	
	less than 1.7%	1.7% to 2.4%	2.5% and over	absolute	relative
10–19	–	2	1	3	2.3
20–29	1	9	7	17	13.1
30–39	–	4	10	14	10.8
Low	1	15	18	34	26.2
40–49	–	3	13	16	12.3
50–59	–	2	16	18	13.8
60–69	1	2	11	14	10.8
Moderate	1	7	40	48	36.9
70–79	3	5	6	14	10.8
80–89	10	12	5	27	20.8
90–99	4	2	1	7	5.4
High	17	19	12	48	36.9
Totals					
absolute	19	41	70	130	100.0
relative	15%	32	54	100	

Small discrepancies in this table are due to rounding.

be catching up, or reducing the gap with the core, in terms of GNP. Despite improvements in infant mortality, life-expectancy, and literacy, there are more people in poverty or facing famine or its possibility than ever before. The gap between the public-health conditions of the core countries and the Third World countries remains enormous. While some countries have progressed in comparison to where they were a decade ago, their ever coming close to core conditions in the foreseeable future seems most unlikely.

Where are the most disadvantaged countries – the "poorest of the poor" – in terms of the criteria herein employed? To qualify for this unfortunate distinction the countries must be in Category I, their annual per capita GNP must be no more than one-fourth the world figure, and the PQLIS must be under 40 or the population growth rate in excess of 2.5 (1.5 times the world rate). This yields a group of 30 countries (Table 8.7) with a total population of 419 million, 8% of the world total. Aside from highly populous Pakistan and Bangladesh, one is concerned here with 28 countries and 218 million people. Twenty of these have less than 10 million inhabitants and 17 have less than 100,000 square miles of territory. Many are therefore rather small countries. The ten or so with sizeable territories have little that is agriculturally usable because of aridity.

Table 8.7 The poorest of us all.

Country	Population† (millions)	Population growth rate	PQLI	Annual per capita GNP
Afghanistan	14.7	2.5	13	★
Bangladesh	101.5	2.8	36	130
Benin	4.0	2.8	25	290
Bhutan	1.4	2.0	26	★
Burkina Faso	6.9	2.6	25	180
Burundi	4.6	2.7	33	240
Central African Republic	2.7	2.4	33	280
Chad	5.2	2.1	27	★
Ethiopia	36.0	2.1	26	140
Equatorial Guinea	0.3	2.2	28	★
Gambia	0.8	2.0	15	290
Ghana	14.3	3.2	44	320
Guinea	6.1	2.4	20	300
Guinea Bissau	0.9	1.9	28	180
Kampuchea	6.2	2.1	38	★
Liberia	2.2	3.1	39	470
Malawi	7.1	3.2	29	210
Mali	7.7	2.8	24	150
Mauritania	1.9	2.9	29	440
Mozambique	13.9	2.8	42	★
Nepal	17.0	2.4	31	170
Niger	6.5	2.8	26	240
North Yemen	6.1	2.7	25	510
Pakistan	99.2	2.7	38	390
Senegal	6.7	3.1	24	440
Sierra Leone	3.6	1.7	11	380
Somalia	6.5	2.6	24	250
South Yemen	2.1	2.9	35	510
Sudan	21.8	2.9	34	400
Togo	3.0	2.8	36	280

★ Under $405 according to the World Bank.
† These countries total 418.8 million people or 8.6% of the world's population. Of this, Bangladesh and Pakistan account for 200.7 million, or 48% of the 418.8 million.

The location of this group of countries is striking (Fig. 8.3). All but eight are in Africa, and, but for four, the African ones form a continuous region which includes all of West Africa except Nigeria and extends eastward to the Indian Ocean and the Red Sea.

The grave difficulties facing the governments and peoples of the Third World are, as should be obvious, not at all entirely of their own making. Their adaptive ingenuity is an important part of the attempt to improve their quality of life. Their integration into the modern industrial world market economy has been accompanied by drastic and increasingly rapid social,

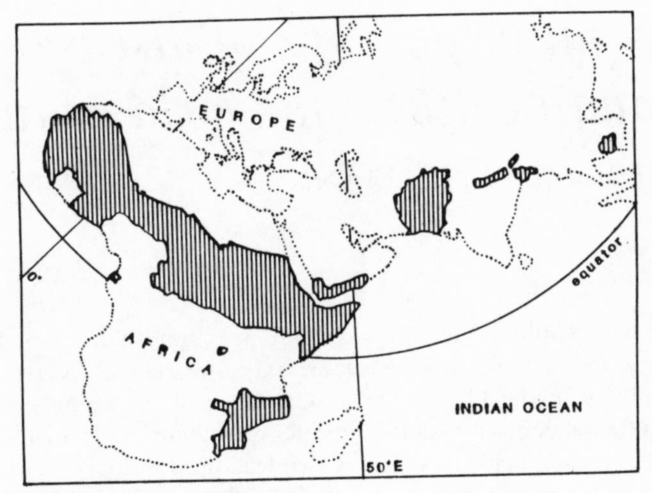

Figure 8.3 The poorest of us all.

demographic, economic, and political changes. The successful completion of this complex process appears to be very problematic. Politico-economic policy questions are beyond the scope of this essay. Our data do, however, point to the need for a recognition of the presence of significant regional differences within the Third World. Country-by-country proposals will not be adequate, nor will general schemes purportedly applicable to the entire Third World.

References and further reading

Harrison, P. 1984. *Inside the Third World.* London: Penguin.
Kidron, M. & R. Segal 1981. *State of the world atlas.* New York: Simon & Schuster.
Kurian, G. T. 1982. *Encyclopedia of the Third World,* revised edn. (3 volumes). New York: Facts on File.
Morris, M. D. 1979. *Measuring the Condition of the World's Poor.* Oxford: Pergamon.
Population Reference Bureau 1975. *World population data sheet 1975.* Washington, DC: Population Reference Bureau.
Population Reference Bureau 1985. *World population data sheet 1985.* Washington, DC: Population Reference Bureau.
UN 1982. *United Nations statistical yearbook 1982.* New York: United Nations.
World Bank 1984. *World tables,* 3rd edn. Baltimore: Johns Hopkins University Press.

9 *Spatial patterns of disease and health: one world or two?*

JEROME D. FELLMANN

Each year over 5 million Third World children, victims of diarrheal diseases, defecate themselves to death. This number is greater than the total number of births annually in the United States, Canada, and the United Kingdom combined. In the poorest sections of the poorest countries half of all children die before reaching their first birthday, while in the developed world cancer and heart disease cause two-thirds of all deaths, and two-thirds of those occur after age 65.

There are indeed at least two worlds of disease and health. One is an affluent world where death rates are low and the chief killers of a mature population are partially avoidable cancers, heart attacks, and strokes. There is as well a second, impoverished, pathogenic world where the deadly scourges of a youthful populace are infectious, respiratory, and parasitic diseases made worse by malnutrition. The appalling disparities between the two encouraged the member countries of the World Health Organization (WHO) to endorse "health for all by the year 2000" as their official target.

Those spatial worlds of disease and health are defined by changing rankings in measures of wellbeing, morbidity, or mortality recorded by the separate countries comprising them. Such rankings are in part anticipated by generalized models that indicate where countries stand on an historical and comparative continuum of health-related development.

Models of development and health

WEALTH AND HEALTH

The richest countries have created environments of health that include adequate food supplies, sanitation and safe water, job-producing industrialization, universal education, and the medical services essential to the wellbeing of their citizens. Such preconditions of health are beyond the present economic grasp of poorer countries. The connection between national wealth and health has both logical and empirical support. Table 9.1 suggests the relationship between gross national product (GNP) per capita and three statistics of wellbeing. Lowest levels of health are reported among African countries, which also record the lowest per capita incomes. Latin

Table 9.1 Wealth and health by major world regions.

Region	Per capita GNP, 1983 (US $)	Life expectancy at birth (years)	Infant mortality rate (per 1,000 live births)	Crude death rate (per 1,000 population)
Less developed	700	58	90	11
Africa	750	50	110	16
Asia	940	60	87	10
Latin America	1,890	65	62	8
More developed	9,380	73	18	9

Source: Population Reference Bureau 1985.

America, with incomes among the highest in the developing world, shows health indicators much closer to those of the developed world. Asia, containing great regional diversity in all health and wealth statistics, occupies a generalized middle ground. By major world regions, developed and developing, the weighted correlation coefficient for per capita GNP and life expectancy is 0.86; for 141 separate countries, it is 0.58. It is therefore argued that a country's ranked position on an economically-based scale of development should as well serve to measure the relative status of health of its population.

It is a persuasive argument that needs to be qualified. Countries with nearly identical per capita GNPs are by no means certain to have comparable morbidity and mortality rates or problems, and countries with low incomes are certainly not automatically unhealthy. Income distribution is far more important in determining the poverty levels so closely associated with disease than is the national average income. With nearly identical 1981 per capita GNPs, the proportion of the population in Paraguay below the absolute poverty level was more than 2.5 times as high as in Jordan; the infant mortality rate was 1.5 times as great (UNICEF 1983). Life expectancies in Sri Lanka, a low-income country, approach those recorded in advanced industrialized countries. Further, the generally high rates of economic growth in the developing countries during the 1950s and 1960s in many instances failed to improve levels of health and wellbeing of the majority of their populations. Because of population growth in the past 30 years, the number of persons in absolute poverty has increased and the number of infant deaths is certainly no less (although infant mortality rates have fallen). Despite apparent correlations, economic indicators alone have proved incapable of categorizing countries on the basis of quality of life or conditions of health of their citizens.

TRANSITION MODELS

The *demographic transition* model traces through changing population characteristics the historical progression of Western countries from conditions of

Table 9.2 Vital rates and expectancies: selected LDCS and MDCS.

Country	Crude birth rate			Crude death rate			Life expectancy at birth (years)		
	1955	1970	1985	1955	1970	1985	1955	1970	1985
Guinea	62	47	47	40	25	23	27	41	40
Madagascar	35	46	45	–	25	17	–	35	50
Nigeria	48	50	48	–	25	17	–	37	50
Tunisia	43	46	33	18	16	10	–	52	61
India	32	43	34	13	17	13	41	–	53
Sri Lanka	38	30	27	11	8	6	57	61	68
Philippines	34	45	32	10	12	7	51	58	64
Thailand	34	43	25	10	10	6	51	57	63
Chile	35	30	24	13	9	6	52	64	70
Colombia	41	41	28	13	9	7	45	61	64
Costa Rica	40	32	31	11	6	4	56	64	73
Guatemala	49	42	43	21	14	8	44	49	59
Japan	19	19	13	8	6	6	66	72	77
UK	15	15	13	12	12	12	70	71	73
USA	25	16	16	9	9	9	70	71	75

Sources: Population Reference Bureau 1985 and UN 1957, 1973.

"underdevelopment" to their present First World status. A first stage of high birth and death rates was succeeded, after 1650, by both an agricultural (Farb & Armelagos 1980) and industrial revolution. Death rates were slashed, birth rates remained high, and explosive growth resulted. In stage three of the transition, urbanization and social change reduced the attractiveness of multiple-child families; birth rates began to fall. Finally, with modern medicine and sanitation in the 20th century, death rates were slashed and in this fourth stage population again stabilized, though at a much higher average age level. In population history terms, a state of "development" was achieved.

There can today be no exact Third World parallels to the transition path traced by Western countries. Conditions are no longer the same. Today, medical advances, pesticides, and improved economic conditions have resulted at least temporarily in plummeting death rates while cultural imperatives still keep birth rates high. Instantaneous communication, international cooperation, and application of modern technologies have jumbled the characteristics and controls of the model's transition stages. Indeed, the model itself has been rejected by some as ethnocentric and inapplicable to cultures other than the European. Nonetheless, evidence of its stages can be discerned in the changing population structures of developing countries.

Table 9.2 suggests how those transition stages can identify individual countries as they pass along the model's prescribed path from demographically-based underdevelopment to full development marked by mature, stable populations. Population data from developing countries are suspect and time series based on them dubious, but we may with caution use them for the trends they suggest. For most of the sample less developed countries (LDCs) shown in Table 9.2, those trends have been toward marked decreases in death rates and increases in life expectancies since 1955; apparently, all of the countries have been beneficiaries of medical technology and economic improvement.

For some, however, such as the subSaharan African countries of Guinea, Madagascar, and Nigeria, there has been relatively little progress along the transition. Continuing high birth rates and high (relatively) death rates indicate a first-stage position. Many of the Asian and Latin American countries show by their rapidly declining death rates and more slowly dropping birth rates that they are passing into the third state. Japan, the United Kingdom, and the United States serve as variant examples of completion of the demographic cycle; Chile and less certainly Costa Rica, Colombia, and El Salvador are following closely behind. India appears at a relative standstill in contrast to neighboring Sri Lanka.

The demographic transition summarizes effects; a parallel *epidemiologic transition* is concerned with causes (Omran 1971, 1977). It begins with the assumption that mortality is a fundamental factor in population change and that mortality patterns alter as a society makes a health-based transition from underdeveloped- to advanced-country status. The beginning Age of Pestilence and Famine is comparable to the first stage of the demographic transition. Infectious, respiratory, and parasitic diseases and malnutrition dominate in which is essentially the pre-modern pattern of disease and health still characteristic of the world's poorest countries. With modernization, the Age of Receding Pandemics is entered, mortality rates begin a progressive decline, and average life expectancy at birth begins to increase from some 30 to about 50 years. A population explosion begins. Gradually, disease and mortality patterns start to shift. Medical and nutritional improvements control epidemics; the incidence of infectious diseases is reduced; and a moderate increase is noted in afflictions of a maturing population. The shift toward those conditions marks the final transition to the Age of Degenerative and Man-Made Diseases. Mortality continues to decline to a low, stable rate; average life expectancy rises to 70-plus years; and fertility becomes the key factor in sustaining population growth.

Because of the parallel nature of the two models, a country's position on the demographic transition should suggest its morbidity and mortality characteristics. At the same time, changing patterns of disease and death should mark clearly a country's progression along a health-determined path to full development in the Western mode. Gesler (1984, p. 10) finds supporting evidence for this proposition. The possibility of *direct* parallels is, however, rejected by Omran (1977, p. 41), who notes "It can be categorically stated that neither the 18th and 19th centuries' experience in the US nor that in Europe is transferable to the less developed countries."

Morbidity and mortality patterns have proved to be more subject to rapidly changing social and technological conditions than to relatively stable environmental circumstances. While this may be encouraging for the control of diseases associated with, for example, tropical regions, it also explains why declining mortality patterns in developing countries have not paralleled the European or American model. In England, for example, the mortality decline during the 19th and part of the 20th centuries was the result of generalized socioeconomic development, including improved nutrition and the creation of water-supply and sanitation systems. Only after the mid-1930s did the application of modern medical technology make a substantial contribution to life expectancy (McKeown 1976, pp. 91–109).

Among developing countries of both Latin America and southern and eastern Asia, the sequence of events has been directly opposite. After World War II very rapid declines in mortality rates and increases in life expectancy occurred immediately after introduction of medical innovations and before there were major improvements in general socioeconomic conditions. Consequently, after a sharp initial improvement, mortality rate decreases slowed substantially (Palloni 1981, Ruzicka & Hansluwka 1982), contrary to the Western model's expectation of continuing gains in life expectancy to recognizable maximum values (UN 1980, p. 31).

DERIVED INDEXES

Nevertheless, the epidemiologic transition model has analytical as well as predictive value. As a society progresses through the transition, degenerative diseases become predominant and an increasing proportion of deaths occur among the no-longer productive elderly rather than among those actually or potentially engaged in economic activity. A measure of "potential years of life lost" (PYLL) has been proposed as a simple calculation that would underscore the social and economic consequences of a reduction in infant and child mortality (Romeder & McWhinnie 1977, cited in Meade 1980, p. 277). Lives lost prematurely – that is, well before some arbitrary but reasonably achievable maximum age is reached – are more costly in economic terms than those lost at or near the established maximum. Child mortality and life expectancy contrasts between the developed and underdeveloped worlds have grim present and future connotations. The PYLL calculation appears to document a country's position on the epidemiologic transition and, by highlighting the significance rather than the cause of death, directs national attention to health programs potentially most beneficial to the wellbeing of the society.

The Physical Quality-of-Life Index (PQLI) seeks to measure how nearly individual countries achieve that wellbeing by meeting the most basic needs of their citizens. Three indicators – infant mortality, life expectancy, and literacy – are each scored 0–100, with 0 an explicitly "worst" performance. A national achievement level is calculated by averaging the three indicators, the result being both a comparative ranking of countries and an index that can show change in national standing over time (Morris 1979). A complement to international comparisons based upon GNP, the PQLI focuses upon results divorced from ethnocentric historical models or from Western concepts of "appropriate" investment strategies. To the extent that the quality of life it seeks to measure reveals as well a national environment of health, the PQLI is still another useful device for international classification and one that sharply contrasts the two worlds of physical wellbeing.

Third World realities

The models and measures reviewed are useful abstractions but they do not address the details of morbidity and mortality in developing countries. By emphasizing theoretical convergences they gloss over the enormity of Third World divergences from patterns of health and wellbeing accepted as the norm in developed countries. The environment of health is a complex web of interrelationships from which isolation of individual elements is difficult and – through undue emphasis – even disruptive. Nonetheless, there is agreement that the achievement of "health for all" must involve bringing to Third World residents First World standards of life expectancy, disease control, infrastructure improvement, nutritional levels, health-care availability, and – to make those changes acceptable and practicable – near-universal literacy. In each of these measures the distance between the poorest-performing LDC and the norms of advanced countries is great; in each, a broad transitional zone of improvement is occupied by an increasing number of countries and a greater percentage of the world population.

Life expectancy is an easily understood statistic of development. Table 9.2 suggests the range that is recorded among countries classified as less developed compared to three "more developed countries" (MDCs): Japan, the United Kingdom, and the United States. In general, the higher a country's average income the higher its life expectancy. Among the richer developing countries longevity approaches or equals the 74-year average in industrialized countries; in low-income countries expectancies may be only 50 years or less.

Infant and child mortality is the major contributor to low life expectancy in developing regions. Some 17% of children in developing countries as a whole and more than 30% in the poorest ones die before their fifth birthday. In industrialized countries the figure is about 2%. The chief culprits are diarrheal diseases and such respiratory infections as influenza and pneumonia. They are rendered more deadly by even moderate malnutrition which increases by a factor of 10 the likelihood that a given child will die from an illness such as measles (UNICEF 1983, p. 21). Together, the deadly trio of diarrheal and respiratory disease and malnutrition accounts for more than half of all child mortalities. Once past the perilous childhood years, adult populations – particularly of urban adults – of developing countries increasingly assume the mortality patterns of their advanced-country counterparts. Indeed, in most instances there is less distinction between the fully urban areas of developing and developed countries than there is between the urban, semi-urban, and rural districts of a single Third World country.

Despite the wide geographical range represented by Third World countries and their separate physical and social environments, certain common disease patterns obtain. First, diseases of highest incidence are those carried in human fecal matter, including intestinal parasites, various infectious diarrheas, typhoid, cholera, and poliomyelitis. They reflect unsafe water supplies and at best rudimentary sanitary disposal practices. Second, as Smith (1979, p. 31) has concluded, "certain diseases are characteristic of poverty and deprivation throughout the tropics": specifically, gastroenteritis and respiratory infections – exacerbated by malnutrition. Third, the traditional vector-borne tropical scourges of malaria, schistosomiasis (bilharzia), trypanosomiasis (sleeping sickness), and river blindness (onchocerciasis), in some instances thought under control or controllable during the 1950s and 1960s, are still serious and growing problems. Fourth, the ultimate pattern of similarity in disease characteristics of developing countries is that of interconnectedness. Whatever the specific disease – localized yellow or dengue fever or cholera, or near-universal tetanus, influenza, or tuberculosis – it is the product of a complex of environmental, economic, and social factors intimately interrelated (Mahler 1981, pp. 7–8).

Characteristic of those sociophysical environments of disease experienced by Third World peoples are (a) the specific disease conditions and health-care opportunities (or their lack) associated with rural and urban-slum residence; (b) the controlling part played by safe water and sanitation; (c) the rôle of malnourishment as the silent accomplice of disease; and (d) the need for a literate population to understand and act upon basic measures of personal and family health.

Within the expanding cities of Third World countries, more than a quarter-billion people live in unsanitary and disease-ridden slums and shanty-towns devoid of adequate water supply or sanitary disposal facilities. The squalor in which they exist provides – through cesspits and water-filled city litter – ideal breeding grounds for carriers of filiariasis (formerly an endemic rural disease) and dengue fever. Yet, despite appalling environmental conditions, it still is healthier to live in tropical Third World cities than countrysides. Most arthropod- or snail-borne infections affect primarily rural folk, who additionally endure the diseases associated with

Table 9.3 Safe water and measures of health in selected nations, 1980–1.

Country	Access to safe water (% of population)	Infant mortality (%)	Life expectancy at birth (years)
Guinea	10	17	44
Madagascar	26	7	46
Nigeria	28	14	48
Tunisia	62	10	58
India	41	12	52
Sri Lanka	22	4	66
Philippines	55	6	63
Thailand	23	6	62
Chile	76	4	66
Colombia	64	6	62
Costa Rica	81	2	70
Guatemala	42	7	59
Japan	98	1	76
UK	99	1	73
USA	99	1	74

Source: Sivard 1983.

unsafe water and poor sanitation and have far less access to even rudimentary health-care clinics than do their urban counterparts.

Safe water and sanitary disposal of human waste are key elements in human health. Their ubiquity in the developed world and their general absence in, particularly, rural areas and urban slums in the Third World constitute a profound contrast in the conditions for health in the two realms. Only one-fifth of the rural populations of 73 predominantly rural countries of Asia and Africa have access to water safe to drink (UNICEF 1983, Table 3). Worldwide, 2 billion people do not have a dependable supply of safe drinking water; three-fourths of the Third World have no sanitary facilities (Sivard 1983). Table 9.3 suggests the importance of access to good water.

Malnutrition is nearly everywhere in the developing world the silent but deadly contributor to disease and death, particularly among infants and children. An estimated one-quarter of the world's children under age five and a total of some 450 million people overall suffer hunger or malnutrition; continued population growth simply compounds the numbers and the problem.

Appalling scenes of famine fill magazines and television screens, but the more insidious and hidden expressions of malnourishment are premature births; low birth weights; abnormally low weight-for-height and weight-for-age ratios; the lethargy, apathy, and low productivity associated with

Table 9.4 Infant mortality and female literacy in selected developing and more developed countries.

Country	Infant mortality 1981 (%)	Female literacy 1980 (%)
Burkina Faso	21	5
Afghanistan	20	6
Yemen Arab Republic	19	2
Chad	15	8
Mexico	5	80
Thailand	5	88
Sri Lanka	4	76
Malaysia	3	61
Hong Kong	1	77
Finland	1	100
USSR	3	98
UK	1	99

Source: UNICEF 1983.

anemia; blindness resulting from vitamin-A deficiency; mental retardation; and a host of other impacts upon physical and mental health. Above all, malnutrition is reckoned to be a contributory cause of between one-third and one-half of all infant and child deaths in developing countries and, coupled with diarrheal dehydration, the direct killer of some 5 to 8 million children each year (Chandler 1985, p. 14).

Malnutrition frequently means more than a lack of calories and proteins; there may be crippling deficiencies in micronutrients and vitamins among, particularly, young children. Nutritional shortfalls and imbalances reflect not only family incomes too low to provide customary food in sufficient volume but also a failure fully to appreciate the rôle of adequate diet in the care of children and the treatment of disease.

Literacy, particularly female literacy, provides the surest guarantee of the eventual achievement of the "health for all" program objectives. It is a particularly valid predictor of infant mortality rates. Numerous studies show that the more educated are mothers, the less likely are their children to die, whatever the family income. The extremes of the accepted relation between female literacy and infant mortality are shown in Table 9.4. A simple regression of infant mortality on female literacy for 119 countries reporting both sets of data for 1980–1 yields a correlation coefficient of −0.87. Although other factors are certainly involved, the association of low infant-mortality rates with high female literacy indicates how important for child health is a mother's awareness of (and willingness to try) recommended health, hygiene, and nutritional practices.

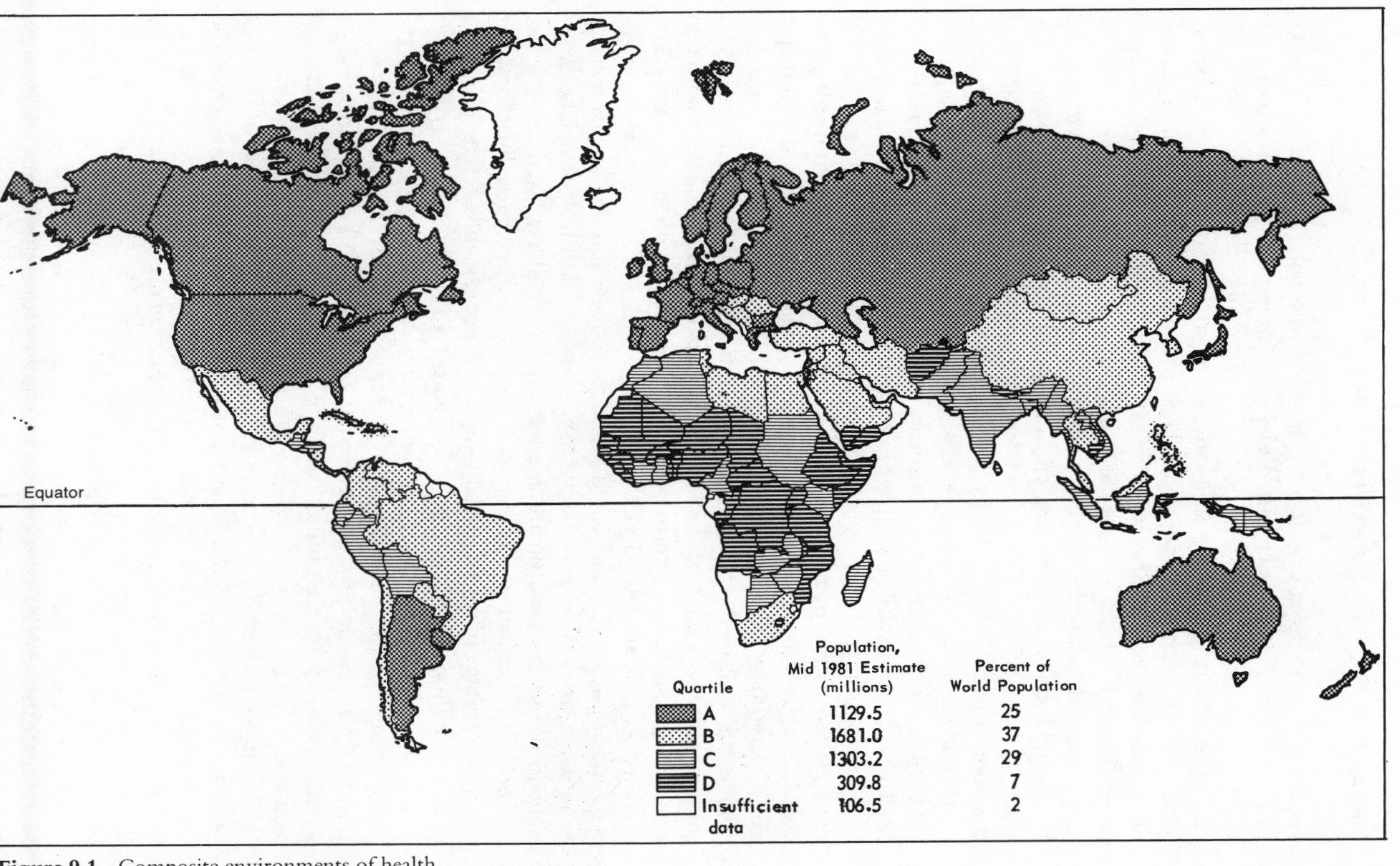

Figure 9.1 Composite environments of health.

Composite environments of health: a world pattern

Health is a generalized condition of bodily and mental wellbeing, a multifactor response to spatially varied physical, biological, and socioeconomic environments. No single component measures its achievement; no single statistic indicates the failure of a country or social group to experience it. There is no mystery about the general determinants of health; equally there is no mistaking the striking world contrasts in the achievement of those determinants by different national units.

At one extreme is the world of affluence and development where good health and long life are the accepted and affordable norms. At the other is the realm of malnutrition, disease, and death occupied by the poorest citizens of the poorest regions of the Third World. In between, there are the bridging worlds where at least some of the components of an environment of health have been achieved and the transition to modern acceptable standards of health has begun. While a simple economic measure of the very rich and the very poor is sufficient to recognize countries at the polar extremes, the multifaceted nature of environments of disease and health demands more detailed analysis to plot the position of countries in transition and to map a global geography of the conditions for health.

Figure 9.1 shows such a mapped classification. It is based upon four measures of wellbeing that show significant spatial variation, are reported in standardized form by international agencies, reveal conditions generally accepted as indicative of environments of health without regard to socioeconomic systems or values, and are independent, additive assessments of progress from less to more developed national health status. The components of the classification are: (a) death rate for children aged 1 to 4 years; (b) accessibility to safe drinking water; (c) presumed average caloric intake as a daily percentage of internationally-agreed standards; and (d) female literacy (Keller & Fillmore 1983, Morris 1979, WHO 1981, p. 34).

The classification displayed on Figure 9.1 is comparative. Countries were ranked from highest to lowest on each of the four factors and the ranks were summed. The resulting scores were the basis for a ranked classification of conditions and potentials of health of the 119 countries for which data were available.

The spatial patterns of the extremes are expected. The advanced countries of East and West – a "First World" of health – scored well on each of the components. The poorest, least developed countries of, particularly, Africa and southern Asia clearly demonstrate the existence of a counterworld of disease and death. What is, perhaps, not expected and not revealed by the spatial distribution is the disparity in the numbers of people who are denizens of each of those worlds of extremes. Only some 7% of global population is represented by the 29 countries of the lowest quartile, while more than one-quarter of the world's peoples are in the most-developed upper quartile.

More than two-thirds of humanity occupies the transitional ground between the two extremes, and more than half of *those* have already advanced beyond the mid-point of the composite national rankings.

The appalling conditions of disease and death afflicting the poorest, least developed countries serve as a continuing reminder that much yet remains before adopted international goals of "health for all by the year 2000" are achieved. Yet, by the multi-factor evaluation of national conditions of health applied here, some two-thirds of the world's population is beginning to approach the health standards set by the most advanced countries. Whatever may be the present divided reality, the prospect is for a converging single world of health.

References

Chandler, W. U. 1985. *Investing in children*. Worldwatch Paper **64**. Washington, DC: Worldwatch Institute.

Farb, P. & G. Armelagos 1980. *Consuming passions: the anthropology of eating*. Boston: Houghton Mifflin.

Gesler, W. M. 1984. *Health care in developing countries*. Washington, DC: Association American Geographers.

Keller, W. & C. M. Fillmore 1983. Prevalence of protein-energy malnutrition. *World Health Statistics Quarterly* **36**, 2, 129–40.

Mahler, H. 1981. The meaning of "health for all by the year 2000." *World Health Forum* **2**, 5–22.

McKeown, T. 1976. *The modern rise of population*. New York: Academic Press.

Meade, M. S. 1980. Potential years of life lost in countries of Southeast Asia. *Social Science and Medicine* **14D**, 277–81.

Morris, M. D. 1979. *Measuring the condition of the world's poor*. New York: Pergamon & Overseas Development Council.

Omran, A. R. 1971. The epidemiologic transition: a theory of the epidemiology of population change. *Milbank Memorial Fund Quarterly* **49**, 509–38.

Omran, A. R. 1977. Epidemiologic transition in the US. *Population Bulletin* **32**, 2–41.

Palloni, A. 1981. Mortality in Latin America: emerging patterns. *Population and Development Review* **7**, 623–49.

Population Reference Bureau 1985. *1985 World Population Data Sheet*. Washington, DC: Population Reference Bureau.

Romeder, J. M. & J. R. McWhinnie 1977. Potential years of life lost between ages 1 and 70: an indicator of premature mortality for health planning. *International Journal of Epidemiology* **6**, 143.

Ruzicka, L. T. & H. Hansluwka 1982. Mortality transition in South and East Asia: technology confronts poverty. *Population and Development Review* **8**, 567–88.

Sivard, R. L. 1983. *World Military and Social Expenditures 1983*. Leesburg, Va.: World Priorities.

Smith, C. E. G. 1979. Major disease problems in the developing world. In *Pharmaceuticals for developing countries*. Washington, DC: National Academy of Sciences.

UN 1957, 1973. *Demographic Yearbook*. Department of International Economic and Social Affairs. New York: United Nations.

UN 1980. *The world population situation in 1979*. Department of International

Economic and Social Affairs, *Population Studies no. 72*. New York: United Nations.
UNICEF (United Nations Children's Fund) 1983. *The state of the world's children 1984*. Oxford: Oxford University Press.
WHO 1981. *Development of indicators for monitoring progress towards health for all by the year 2000*. Geneva: World Health Organization.

10 *The status of women in the Third World*

ALICE C. ANDREWS

In recent decades there has been much discussion, even controversy, about the "status of women." What does the term mean, how is it measured, how does it vary in different parts of the world, and what, ideally, should it be? The answers to these questions, particularly the last one, vary cross-culturally. The status of women, or even the perception of the ideal status of women, does not correspond neatly to stratifications of First, Second, and Third Worlds, nor to the standard culture realms of regional geographers. It varies between them, but also within them. Factors of religion, education, socioeconomic standing, and political ideology color our thoughts about women's rôle in society.

The United Nations designated 1975–85 as the UN Decade for Women, a period in which governments were urged to move from broad guarantees of gender equality toward more specific programs that promoted such equality and corrected past discriminatory practices. In 1981 the capstone Convention on the Elimination of All Forms of Discrimination Against Women entered into force. By the time of the Nairobi Conference that brought the Decade to a close, over 60 member nations had ratified this UN Convention, which ensures civil, political, educational, employment, health, and marriage rights for women.

The Decade for Women has seen important progress, some of it embodied in law and some in heightened awareness of the problems and needs of women. This chapter will first summarize current thought about what is meant by women's status; it will look at various indicators of that status; and it will then give a region-by-region description of the present condition of Third World women, using data sources that became available in the flurry of activity that accompanied the Nairobi Conference.

Status

Any attempt at a worldwide stratification of the status of women must come to grips with the terms involved. The "status of women" has become a catchphrase, but its meaning is far from clear. "Status" is defined in standard dictionaries as meaning condition or position with regard to law; it may also mean rank or standing, prestige, or simply state or condition. In assessing the status of women, all these meanings should be considered.

To understand the status of women in a particular region, we need to know about their status in the eyes of the law (is it equal to that of men?), their prestige (do they have opportunities to gain prestige socially, economically, and politically?), and their overall condition (what is their quality of life, how is their wellbeing?). To these basic meanings the terms "independence" and "access to resources" should be added. Ideally women should have some independence and autonomy in their lives, as do men. They should have an opportunity to participate in making decisions, not simply have them made for them. As a matter of social justice and human rights, they should have equal access to the resources of a society. Most development theorists now believe that women should fully participate in development and decision-making at all levels – family, community, and national.

Perhaps the most comprehensive and useful definition of women's status has been put forward by Ruth Dixon, a pioneer in the literature on women in development. Shortened and simplified, it states that the status of women is the degree of women's access to and control over material resources (food, land, income, and other forms of wealth) and social resources (knowledge, power, and prestige) within the family, in the community, and in the society at large (Dixon 1978).

Indicators of women's status

Various demographic and economic statistics have been used as indicators of the status of women. Two of the most complete recent sources of such indicators are the *Women of the world* series (US Bureau of the Census 1984a & b, 1985a–c) and *Women: a world survey* (Sivard 1985). Other sources of indicators published near the time of the Nairobi Conference were wallcharts produced by the Population Reference Bureau and the International Planned Parenthood Federation. Regular, continuing coverage of many indicators is provided by the United Nations and by the World Bank.

A major problem in the use of indicators to measure women's status is the difficulty of securing comparable data. Incomplete and faulty registration, the use of different definitions and categories, and different time frames pose problems. The reporting of women's participation in the labor force, for example, varies drastically from country to country.

Statistics regularly reported to the UN include literacy, enrollment in education, fertility, and life expectancy. Some geographic researchers have tried to combine a number of indicators into a single index that could be mapped (Lee & Schulz 1982, Andrews 1982). There is as yet little agreement as to what indicators should be used in such an index, even if reliable, complete, and comparable data were available. Over all looms the difficulty of cross-cultural comparison. A composite index might be effectively used within a large country or within a region that is reasonably homogeneous, but to apply it worldwide poses enormous difficulties.

Table 10.1 contains a sampling of indicators for selected countries in four different Third World regions. For comparative purposes indicators from a smaller number of First and Second World countries are included.

LITERACY

There is a huge range in female literacy from the very low rates of the poorest Third World countries to almost universal literacy in the First and Second Worlds. The discrepancy between male and female rates is greatest in Africa but is still considerable in parts of the Middle East and in the Indian Subcontinent.

SECONDARY SCHOOL ENROLLMENT

The lowest proportions of females enrolled in secondary school are found in the Sahel of Africa, but similarly low rates for females and even greater disparities between males and females are found in Bangladesh and Pakistan.

LABOR FORCE

The percentage of women aged 15–64 in the labor force would be a useful indicator but is unreliable, due to differences in definitions and reporting practices. For example, the same percentage appears for the United States and Ethiopia, while other indicators show relatively high status in the US and very low status in Ethiopia. Despite these data deficiencies, the very low rates in many Moslem countries are noteworthy.

MARRIAGE

The percentage of women aged 15–19 who are married seems to correspond with other indicators. It is quite high in countries where women's levels of education are low. In Bangladesh, notorious for child brides, 70% of women are married before they are 20. The rate is lowest in Northern Europe and in East Asia; in the latter it is very low in both capitalist Japan and communist China.

FERTILITY

The total fertility rate (number of children who would be born per woman during her reproductive life span if current levels of reproduction prevailed) reaches an amazing eight children in Kenya. It is generally high in Africa, the Middle East, and Pakistan and Bangladesh. It is lower in Latin America, but varies with level of development, being highest (over six) in Guatemala, Honduras, and Bolivia. Communist Cuba has a low rate, identical to that of the United States.

Table 10.1 Some usual indicators of the status of women in selected countries.

	Literacy/education		Labor force	Marriage		Fertility		Mortality	Participation in government
	Adults literate: male/female* (%)	Enrollment in secondary school: male/female† (%)	Women aged 15–64 in labor force* (%)	Legal age: male/female†	Women aged 15–19 married* (%)	Total fertility rate*	Births per 1,000 women aged 15–19†	Life expectancy at birth: male/female‡ (years)	Women members in legislature/ cabinet§ (%)
Burkina Faso	15/3	4/2	81	–	54	6.5	218.3	43/46	–/4
Ghana	43/18	44/27	56	21/21	31	6.5	206.3	53/57	–/28
Ethiopia	8/1	16/8	60	20/15	61	6.7	206.1	45/49	–/–
Kenya	60/35	23/15	44	18/18	28	8.0	238.7	55/59	2/0
Cameroon	68/45	25/13	40	18/15	53	6.5	184.6	52/55	14/15
Zambia	84/67	21/11	41	16/16	–	6.8	207.6	49/52	3/–
Iran	48/24	54/35	14	15/13	34	5.6	88.5	60/60	–/–
Egypt	54/22	64/39	6	18/16	22	5.3	76.3	56/59	8/3
Algeria	57/32	42/29	4	18/16	–	7.0	111.9	55/59	4/–
Turkey	83/53	57/28	45	17/15	22	5.1	67.0	61/66	3/0
Syria	60/20	59/37	10	–	23	7.3	111.9	65/69	–/–
China	79/51	53/35	55	22/20	4	2.1	0.0	65/69	21/5
India	48/19	39/20	44	21/18	57	4.5	40.8	55/54	7/–
Bangladesh	37/13	24/6	12	18/16	70	6.4	195.1	48/49	–/–
Pakistan	36/15	27/7	12	18/16	31	6.4	91.0	51/54	–/–
Philippines	84/81	58/68	52	16/14	14	4.8	40.8	62/66	4/7
Indonesia	78/58	36/24	33	19/16	30	4.4	31.2	52/55	9/–

Cuba	96/95	72/77	38	16/14	30	1.8	67.9	73/77	23/8
Mexico	89/84	54/49	31	16/14	19	4.4	99.0	64/68	33/–
Bolivia	76/51	37/31	23	14/12	17	6.3	90.3	49/53	2/0
Brazil	78/74	29/35	31	18/16	18	4.0	57.4	62/66	–/–
Venezuela	80/73	41/38	31	14/12	20	4.1	95.2	65/71	5/10
Chile	90/88	53/62	30	14/12	10	2.6	61.6	68/72	4/–
US	99/99	94/96	60	18/17	8	1.8	63.0	71/78	5/15
Japan	99/99	92/93	54	18/16	1	1.8	3.3	74/79	3/0
Sweden	99/99	80/90	81	18/18	1	1.6	22.9	75/80	29/25
France	99/99	77/93	53	18/15	5	1.8	23.9	71/79	5/15
USSR	99/98	21/22	71	18/18	–	2.4	33.9	65/74	33/0
Bulgaria	96/93	83/82	67	18/18	18	2.0	74.6	70/75	22/4
Czechoslovakia	99/99	34/58	73	18/18	8	2.1	42.7	68/75	28/–
Yugoslavia	92/76	85/79	48	18/18	16	2.1	53.8	69/74	15/3

Sources
* PRB 1985. † IPPF 1985. ‡ World Bank 1984. § Sivard 1985.

Fertility rates for teenage women (number of live births per 1,000 women aged 15–19) may be a particularly sensitive indicator of the status of women; births to very young women, as well as to those in advanced age brackets, pose health hazards to both mothers and babies. Here again Kenya leads the list with an age-specific fertility rate of almost 240. This roughly translates to one in four teenage girls having a baby in a given year. For much of Africa, and for Bangladesh, the rate is around 200. At the other end of the spectrum, China reports a rate of zero and Japan a rate of only three.

MORTALITY

The most generally available indicator of mortality is life expectancy, although more specific indicators such as maternal mortality and sex-differentiated infant mortality rates would be more sensitive measures of female status. Throughout most of the world mortality rates favor females, for reasons that appear to be biological although they are not entirely clear. This results in the well-known fact that women have longer life expectancies than men. The male/female differential is most pronounced in the highly developed countries, is significant in Latin America, and is apparent even in Africa, despite high mortality rates for both sexes. The lower life expectancy for women in some South Asian countries is thus unusual and provides evidence of low female status.

Regional summaries for developing areas

The US Census Bureau, in cooperation with the Office of Women in Development, USAID, in 1979 began compiling information about women in developing areas (the "Women in Development Data Base"). Two regional reports were issued in 1984 and two in 1985, at the end of the UN Decade on Women. The following regional descriptions are based largely on these data summaries.

LATIN AMERICAN AND THE CARIBBEAN

Indicators place the status of Latin American women at a higher level than that of their sisters in Africa and Asia. Latin American countries have higher per capita GNPs and are more modernized. Although the indicators, particularly educational ones, are higher, the disparity between males and females is present. One demographic characteristic in which Latin American women are markedly different from Asian and African is their migration behavior. In Latin America, women outnumber men in the rural-to-urban migration streams, resulting in high sex ratios (the number of males per 100 females) in the countryside and low ones in the cities.

Levels of literacy and school enrollment have increased in recent years, and the discrepancy between male and female rates has decreased. Literacy rates are high in urban areas, with a median regional percentage of 84 for women and 91 for men. School enrollment rates for both sexes are reasonably high, ranging from 60% in rural areas to almost 90% in cities, and the gender differences in enrollment are minimal; in some cases female enrollment actually exceeds that of males at primary and secondary levels. The median figures mask striking intra-regional differences; high enrollments are found in Argentina, Chile and Uruguay (the Southern Cone) and in some small Caribbean countries, and lows are found in Haiti and Central America. The gender gap in literacy is now apparent only in the older age brackets and disappears at younger ages, due to significant improvement in educational opportunities for women in recent years. Female education is more nearly equal to that of males in urban areas, as opposed to rural.

At higher levels of education, discrepancies between the sexes still remain. Women are concentrated in traditionally female fields such as education. Dentistry, laboratory technology, and pharmacology have also become overwhelmingly female fields. In recent years, however, more women are entering sex-neutral fields – computer science, statistics, advertising, journalism, and the social sciences. There are very few women students in the sciences or in business. Women studying medicine are clustered in "appropriate" specialties – gynecology, obstetrics, and pediatrics.

The overall participation of women in the labor force is difficult to assess, because so many of their work activities do not fall into the categories of the formal labor force. Women have always been involved in the informal labor market in Latin America; there is now more movement into paid employment in the formal sector. This trend is tied in with rapid urbanization and the predominance of women in the rural-to-urban migration pattern. In the cities, the largest proportion of women is found in the service sector. Many work as domestics or in other menial, poorly-paid jobs. In rural areas, women continue to work in agriculture, caring for crops and animals and marketing farm products, in addition to their household and childcare duties. They are thus not represented in the cash economy or in the labor-force statistics.

SUBSAHARAN AFRICA

By almost any measure, the people of subSaharan Africa are disadvantaged. This is perhaps particularly true of women. In this region, the question of women's access to resources becomes not only a matter of social justice and human rights but also an important and well-publicized development issue. Women are the primary food producers in the region; they are also engaged in small business, especially the marketing and distribution of food and household items. In the past attitudes about gender rôles have downplayed the importance of women's work. New development theorists stress the need for educating women and involving them in planning.

The subSaharan region is characterized by high fertility, rapid population growth, and a heavy youth dependency burden per adult worker. The median proportion under age 15 is 45% for the region as a whole. The region is also heavily rural. Migration streams are usually sex-selective for males, who seek work in cities and mining areas. The result is that for the prime working ages (15–49), there are higher percentages of women in rural areas and of men in urban areas.

Literacy rates, which vary greatly from country to country, favor males. There is evidence of improvement in women's rates in the past two decades, and the female disadvantage decreases in the younger age groups. The poorest rates of overall literacy, as well as some of the most marked male/female discrepancies, are found in the Sahelian countries, with Ethiopia as the worst case. Literacy improves to some degree in West Africa and even more so farther south. The same general pattern of improvement towards the south is confirmed by figures for school enrollment.

There has been a tendency to exclude girls from new post-secondary programs in agriculture and commerce, but these are fields in which women need training, for traditionally many of these activities have been in their hands. At post-secondary levels, single-sex institutions and tracks are common, with females usually relegated to studies such as nursing, food preparation, and beauty culture.

The modal minimum legal age for marriage in the region is 16 for women and 18 for men. Marriage rates are very high; marriage is the accepted state in subSaharan Africa. Women marry at younger ages than men and rural women marry at younger ages than urban dwellers. The age difference between husband and wife is usually 5–10 years. Polygamy is still common, and older men often take much younger women as second wives. Jeanne S. Newman, who wrote the textual analysis for the Saharan volume of *Women of the world* (US Bureau of the Census 1984b), aptly points out that the family is the fundamental social institution in the region; marriage is a basic way through which family ensures its stability and wellbeing. Women achieve status within the family; thus marital status is a critical variable in understanding the status of women in this region.

NORTH AFRICA AND THE MIDDLE EAST

An outstanding demographic characteristic of this region is a high migration rate – both internal, from the countryside to the cities, and international, from labor-exporting to labor-importing countries. The rural-to-urban movement is dominated by males, resulting in a heavy proportion of women in the working ages remaining in rural areas. The international stream is even more sex-selective for males, leaving the working-age population of labor-exporting countries such as Algeria heavily skewed towards females, while the opposite is true in labor-importing countries like Saudi Arabia.

Literacy rates vary greatly within the region but in all cases favor men. "Worst cases," in terms of both overall rates and the gender gap, are Yemen (only 2% of women literate, compared to 25% of men), and Afghanistan (9% of women literate, 37% of men). Saudi Arabia's rates of literacy are also low (35% for men, only 12% for women). Literacy rates for Syrian men are three times as high as those for women (60/20) and for Libyan men four times as high (61/15). On the other hand, rates of between 70% and 80% for men compared with rates in the 50% range for women are found in Jordan, Kuwait, and Lebanon. The discrepancy between males and females is somewhat reduced in another educational indicator, the percentage enrolled in secondary school. These figures average around 43/31 for North African countries and about 52/33 for the Middle East. The gender gap for all educational indicators is greater in rural areas than in urban.

It is particularly difficult to assess the economic status of women in this region. The percentage of women in the labor force is very low, averaging about 8% for the region as a whole. It is less than 10% for many countries; rises to around 20% in Kuwait, Bahrain, and Lebanon; and reaches totally atypical highs in the 40% range in Cyprus, Israel, and Turkey, the most Europeanized countries. Based on the relatively few countries reporting such information, the percentage of women who are "economically active" is higher (15%) in rural than in urban areas (10%), but 60% of these economically active women in rural areas are reported as being unpaid family workers. The net result seems to be that very few women are gainfully employed. The occupations in the modern sector that are most available to women are the very standard female ones of nurse, teacher, and secretary.

Fertility rates in the region hover between five and six children; they are slightly lower than those for subSaharan Africa and slightly higher than rates for most Asian countries. Outstandingly high fertility rates are found in North Yemen, Saudi Arabia, Syria, and Libya. Rather low rates are found in Lebanon and Cyprus. It is tempting to equate high fertility in the region with Islam and to note that some of the highest rates are in countries where Islamic fundamentalism is strong. This might be too simplistic, however, for studies have shown that women's education interacts with religion. For example, in Lebanon, differences in fertility between Moslem, Christian, and Druze women were significant at low levels of education and not at all significant for women with higher educational levels.

Mortality data for the region are poor, with perhaps only half of deaths being properly registered. Figures in the "Women in Development" data base are often based on sample surveys. With these caveats, it is possible to make some observations from the mortality data reported for this region. Life expectancy is typically higher for females than males in the region, but the difference is small. Life expectancies in Afghanistan and Iran were lower for women than men, in 1979, probably due to high maternal mortality. Infant mortality was higher for girls than for boys in Iran and Jordan. The proportion of children dying before their fifth birthday was higher for girls than boys in Iran, Syria, and Egypt. All these are departures from the general worldwide pattern of higher mortality for males than females at all ages and are suggestive of low female status.

ASIA

Asia is a huge continent with high population clusters. Well over half the world's women live in Asia, excluding the Soviet Union. Almost one billion of the world's women live in China and the Indian subcontinent. The situation of women in Asia varies greatly from communist China to capitalist South Korea, from Hindu India to Buddhist Thailand, but, with the exception of Japan, the countries are all Third World.

Rates of literacy and educational enrollment are higher for men than women throughout the region, but the gender gap in education varies from a high in Nepal to very little in the Philippines. Literacy is higher among younger women as a result of recent improvements in educational opportunities; it is also higher among urban than rural women. The school enrollment differential is small for the youngest age group but increases with age.

Comparing rates of economic activity is difficult, as usual; much of women's work is not reported and statistical practices vary from country to country. Some generalizations, however, can be made. There is a discrepancy between reported rates of labor-force participation between women and men. Women's rates are lowest in the Indian subcontinent, where they are especially low in Pakistan and Bangladesh and only slightly higher in India and Nepal. They are higher in East and Southeast Asia, particularly in China and Thailand. Most female occupations are in agriculture, sales, and services. Higher rates of female economic activity are found in rural areas. Women's labor-force activities are in response to poverty and are added to the traditional duties of wife and mother. They add responsibility, hours of work, and some income to the family. For the most part they do not add to the status of the individual woman or to the income she has at her own disposal.

Both the legal minimum and the reported mean age at marriage are lower for women than men. Both sexes marry earlier in rural areas. Women still gain status through bearing children, especially sons. The mother of sons gains prestige and authority, especially authority over her daughters-in-law. This is particularly true in the Indian subcontinent, where fertility rates remain high. In China, fertility is much lower (a total fertility rate of 2.1) as the result of the rigid population-control policy and the much-publicized one-child family policy. It is also low (2.0) in South Korea. Fertility rates in Southeast Asia tend to be higher than East Asia and lower than the Indian subcontinent. In the latter region, fertility remains relatively high, with a total fertility rate of over 6 in both Pakistan and Bangladesh. India has only managed to pull its rate down to 4.5, after over 30 years of an antinatalist population policy.

In this same part of Asia, the Indian subcontinent, mortality rates appear to be important indicators of the low status of women. Female life expectancy at birth is lower than male in India, Pakistan, and Nepal. The emphasis on sons seems to result in adverse effects on female mortality. It has been suggested that the explanation for this includes differential health care and nutrition, as well as the sociopsychological attention given to male children. In other words, boys are favored over girls in the provision of food, medical care, and infant attention. A much-publicized study done in India in the early 1980s also showed that when women were able to obtain prenatal gender determination, an overwhelming proportion decided to abort female fetuses.

Conclusions

By our definition of status and from the evidence of demographic indicators, it is clear that the status of women is poor in the Third World and that it varies greatly. For purposes of comparison, it is worth remembering that in the First World women are still not satisfied with their status; the struggle for equal rights continues. That struggle, however, is for greater recognition, equality, and power in largely economic and political realms, not in basic human rights. In the communist Second World, all governments officially endorse the principle of equal rights for men and women; girls have equal access to education, at least up to higher levels. But, there is something that in Russia is simply referred to as "the problem of women," and that is the fact that women are particularly subject to the total load of household work in addition to full-time employment (Goodman 1985). It is also clear that women are reasonably well represented in communist legislatures, but not in proportion to their number in the population; they are never found in top executive jobs.

In the Third World, the status of women is lower and the problems are great. Indeed, the status ranges from poor to abysmal. Third World women work long hours. Paid work outside the home is an addition to the traditional burden of household work, child care, and often agricultural production. Women bear many family-related responsibilities. They are responsible for fertility and fidelity in marriage, for child care, for family health and nutrition, and generally also for education of children. They are held accountable for infertility or for producing daughters instead of sons; they can be divorced for this failure. At present, most of them are struggling, not for equal pay or equal political power, but for simple human and civil rights and for access to food, health care, and paid employment.

Within the Third World there are striking variations in women's status, some due to custom, some to law, some to religion, some simply to poverty and lack of development. The highest status, by the indicators examined, is found in Latin America, which is also farthest along on the road to economic development and is the most Europeanized region of the Third World. Within Latin America, the indicators are higher in the most Europeanized countries and in communist Cuba.

In subSaharan Africa the status of women is very poor, with high fertility being one of the major indicators. Some of the most serious questions arise in North Africa and the Middle East and in the Indian subcontinent. In these areas some sensitive indicators of wellbeing, such as length of life and infant mortality, seem to indicate that female life is not as highly valued as male and that women suffer from discriminatory practices. It is impossible to note these problems without asking questions about religion. Hinduism certainly cannot be said to promote high female status. Customs such as suttee were repressed by the British only with great difficulty, and problems of bride-burning still persist. Only recently the persecution of women during the festival of Holi was highlighted in newspapers, which pointed out that during the festival women cannot go on the street or ride public transportation without fear of molestation. Even though Mrs. Gandhi was a powerful and revered leader, the rest of her sex have a long way to go.

In Islamic countries, especially those dominated by fundamentalism, the problems of women are more institutionalized. Islamic law, the Shari'a, allows women to inherit property (in smaller shares than males) and to receive a financial settlement or marriage gift. These provisions protect women to some degree, but they are counterbalanced by customs such as polygamy (although it is not now widely practiced) and by unequal divorce laws whereby a man may divorce his wife by renunciation. Other customs that work against women are early marriage and segregation of the sexes. In practice these things vary greatly, from liberal states like Tunisia to ultra-conservative Saudi Arabia and fundamentalist Iran. After the switch to Islamic government in Iran, education about birth control was stopped, large families were encouraged, the minimum age for marriage was lowered to 13 for females and 15 for males, and changes in marriage law reduced women's rights (*People* 1983, p. 28).

If there is a single word that sums up the hope for improvement in the status of women in the Third World, it is education. This is emphasized by the fact that the World Fertility Survey showed that fertility and child health are even more closely related to women's ability to read and write than they are to income. Improvements are being made. Rates of school attendance for girls are increasing in most parts of the Third World. There are still great needs ranging from the simple need for improvement in basic literacy in some areas (the Sahel), to a need for more opportunities in technical education. In the Third World, women's entry into the labor market is usually a response to poverty. Their opportunities for meaningful employment can be vastly enhanced by opening educational doors.

References and further reading

Andrews, A. C. 1982. Toward a status-of-women index. *Professional Geographer* **34**, 24–31.

Dixon, R. 1978. *Rural women at work*. Baltimore: Johns Hopkins University Press.

Goodman, E. 1985. Red Squares. *Washington Post*, May 14, 1985, A19.

Henshall Momsen, J. and J. G. Townsend 1987. *Geography of gender in the Third World*. London: Century Hutchinson & Albany, NY: State University of New York Press.

IPPF 1985. *Youth in society: 1985 IPPF/people wallchart*. London: International Planned Parenthood Federation.

Lee, D. & R. Schultz 1985. Regional patterns of female status in the United States. *Professional Geographer* **34**, 32–41.

People 1983. Iran turns back the clock. *People* (publication of the International Planned Parenthood Federation) **10**, 28.

PRB 1985. *The world's women: a profile*. Washington: Population Reference Bureau. (Wallchart.)

Sivard, R. L. 1985. *Women: a world survey*. Washington: World Priorities.

US Bureau of the Census 1984a. "Women of the world" series: *WID–1: Latin America and the Caribbean*. Washington: Department of Commerce.

US Bureau of the Census 1984b. "Women of the world" series: *WID–2: subSaharan Africa*. Washington: Department of Commerce.

US Bureau of the Census 1985a. "Women of the world" series: *WID–3: Near East and North Africa*. Washington: Department of Commerce.

US Bureau of the Census 1985b. "Women of the world" series: *WID–4: Asia and the Pacific*. Washington: Department of Commerce.

US Bureau of the Census 1985c. "Women of the world" series: *WID–5: A chartbook for developing regions*. Washington: Department of Commerce.

World Bank 1984. *World development report 1984*. New York: Oxford University Press.

11 *Toward a geography of freedom*

LAURENCE GRAMBOW WOLF

The governance of any Third World country presents serious problems. Quite aside from adapting traditional ideologies to an apparently inescapable process of modernization, there are structural politico-economic conflicts which often defy peaceful resolution. There is the conflict between the private business interests and the state bureaucracy. These two classes are not in conflict at all points and times, but there is a basic opposition over the economic rôle of the state within a capitalist frame of reference. There is a conflict between either or both of these and the peasants and workers . . . the old class struggle of profits versus wages. There is the conflict between countryside and city, which can result in the impoverishment of the peasantry in order to keep food prices low for all urban classes and to advance export crops at the expense of local food production. There is the conflict between the military and the civilian authorities in determining the course of events, and there can be ethnic and religious conflicts as well. Not all of these conflicts bedevil the polity of all Third World countries, of course, and each country has its own variation of this set of problems.

Beyond all this domestic stress lie the relations with the rest of the world, which usually involve relations with the advanced industrialized countries, the transnational corporations, and the agencies of international finance, rather than relations with other Third World countries. There is then a built-in conflict between the need of the domestic ruling class to cooperate with (to subordinate itself to) these foreign interests and, on the other hand, the need to pursue its own material interests.

Further to this is the fact that many Third World countries are very small and poor in resources, and would have difficulty creating a modern infrastructure under the best of circumstances. We then have the basis for understanding why the conditions of political freedom occur but rarely. The development of a tradition of civil liberty and peaceful, democratic political decision-making is facilitated if the economic pie is usually growing. This has not been the case in most of the Third World.

Third World polities

A great variety of political regimes exist within the Third World. A comprehensive classification is not within the scope of this study, but salient types and examples can be noted. There is, for instance, still an Arabic form of absolutist monarchy in Saudi Arabia, with some variations on that theme elsewhere. There are still old-fashioned dictatorships that rely on corruption and suppression of dissent without the use of modern ideologies, as in Paraguay, and, until recently, Haiti. There is theocracy resurgent, as in Iran. Mexico has a semi-democracy in which political life is all but monopolized by one party. Several countries have varieties of leninist marxism, as in South Yemen, Vietnam, and Cuba. A common type of polity is that in which the national military establishment assumes the rôle of "protector of society." This results in a regime that attempts capitalist economic development with more or less state interference in the economy. Traditional Western democratic governmental forms then function as window-dressing or as a stop-gap reluctantly resorted to at times when openly military rule appears politically inauspicious. Prime examples are Argentina (which had military coups in 1930, 1943, 1944, 1945, 1955, 1962, 1966, and 1976), Brazil, and Turkey, though others could be mentioned.

The Free World and the Third World

It is common practice among our politicians, journalists, commentators, and even many scholars, to contrast the Free World to the Communist World without much qualification of the first term. If the term "Free World" is to have any real substance, it should not be applied loosely, lest it be inferred that Uganda under Idi Amin was free, or Haiti under the Duvaliers.

A widely accepted description of a politically free society is the traditional "bourgeois democratic" one; to wit, a society in which freedom of speech, press, assembly, and religion are traditionally accepted ideas; in which two or more political parties contest electorally for office; in which the transfer of power from one party or coalition to another is peaceful; and in which, subject to all the usual qualifications about shortcomings, limitations, and violations, these characteristics are the accepted dominant tradition for national and local politics. Many trenchant and legitimate criticisms of such polities have been made, yet the difference between a polity with and a polity without these freedoms is often a matter of life or death for political activists. What needs to be kept in mind all the while is that these polities are not absolute in their freedom, but relative; that they are not fixed in time, as struggles for the rights of the unpropertied, of women, and of racial and ethnic minorities attest; and that freedom is not a mystical abstraction: it is

Table 11.1 States with a long, uninterrupted democratic tradition.

Great Britain*	Iceland	Switzerland
Ireland	Norway†	Canada
The Netherlands†	Sweden	New Zealand
Belgium†	Denmark†	Australia
Luxembourg†	Finland	USA
		(France)‡

* Great Britain, not the United Kingdom, as Northern Ireland's prolonged internecine strife disqualifies it.

† Wartime interruptions due to Nazi occupation are not counted.

‡ Which table (11.1 or 11.2) is most appropriate in this case depends on one's evaluation of the ease with which France acceded to the Nazis, and the strength of traditional antidemocratic elements in French politics.

Table 11.2 Democratic states with a less firm tradition.

West Germany*	Portugal*	Costa Rica§
Austria*	Greece*	India¶
Italy*	Japan*	The Philippines*
Spain*	Israel†	
France‡	Jamaica†	

* These states have had periods of democratic rule interrupted by indigenous dictatorial regimes.

† Israel is the most outstanding example of states which have not been in existence long enough for the strength of democracy as the *traditional* polity to be evaluated with assurance. Jamaica is cited as an example of several small former colonies with democratic polities which have not been independent long enough to have developed a firm political *tradition*.

‡ See footnote ‡, Table 11.1.

§ This is the best example in Latin America: its latest civil war occurred in 1948. Until recently, Uruguay was the most firmly democratic Latin American state. After a dictatorial interlude, democracy has been restored there.

¶ The peculiarities of Indian society and its polity make it in some respects a borderline case.

best defined operationally, for freedom on paper, legally guaranteed, means little unless in actual practice these legal rights can be exercised.

A very limited number of countries qualify for inclusion in the Free World so defined (Table 11.1). Others can be considered which conform currently but have had dictatorships some time during the present century (Table 11.2). Beyond these two categories, a democratic polity becomes an ephemeral or dubious matter or one not sufficiently long-lived for a firm evaluation to be made.

The list of countries in the Free World could be considerably extended if one were to include all of those which have the outward trappings of a democratic polity in at least some rudimentary form at the moment, but the position taken in this study is that a span of time of at least two (and preferably more) generations must be considered. To consider as free states those in which democracy can be turned on and off, or in which it is merely tolerated as a (perhaps temporary) expedient, is to indulge in a superficiality which can serve propaganda purposes better than those of genuine understanding.

There are now some 11 or 12 Third World countries which are marxist: the remainder are capitalistic. Of the latter, a few are in our second category of politically free countries and none are in the first category. Obviously, capitalism and democracy cannot be equated. The overwhelming majority of Third World capitalist countries do not permit their citizens to express their opinions freely or to undertake political activities without fear of harassment or something far worse.

A survey of political conditions by Amnesty International reveals that in at least 15 clearly non-marxist Third World countries political dissidents, suspected political dissidents, and even relatives of actual or suspected political dissidents have been "disappeared." Concentration camps, as found in Nazi Germany or under Stalin's rule, are becoming obsolete. It is much more efficient to have agents of the government simply abduct their victims, who are then murdered secretly, or, in some cases, released whenever it suits the convenience of their tormentors. In 1983, in more than 40 Third World non-marxist countries, numerous cases were discovered of persons apprehended for publishing material critical of an official or of the regime, or for some other form of nonviolent dissent. Arbitrary arrest, detention (often incommunicado and sometimes long-term) without legal charges being made, and the torture of prisoners occur in a distressingly long list of countries, even if those experiencing a civil war or an attempted coup are ignored. The fact that these acts of tyranny may affect only a small number of persons (tens or hundreds) does not make them less important. Political activity usually engages only a small percentage of any populace.

Turkey, for instance, can hardly be considered part of the Free World when the military determine which political parties may participate in an election, when prominent citizens are placed in Turkey's notorious prisons for advocating peace, and when newspapers can be shut down for displeasing the military. In Nepal and Mauritania all political party activity is prohibited. In South Africa the lack of political rights for the majority of its inhabitants is an international *cause celèbre*. The political shortcomings of marxist governments are quite well known. The fragility or the total absence of political freedom in most non-marxist countries is not made explicit in any systematic, sustained way and therefore is not so clearly part of the general public's political awareness.

The ideas that political freedom can be exported to Third World countries and that the only evidence one needs of a successful export is the occurrence of elections (no matter how rigged, no matter what hazards dissenters must endure), and a parliament or congress (no matter how unrepresentative), do not survive even cursory scrutiny. It may be that the stressful political and social conflicts that arise as a country is integrated into the world economy are too great for a free polity to develop based on a firm, dominant, traditional democratic consensus, or even to occur as more than a partial and temporary condition.

The traditional "bourgeois democratic" model of political freedom has not been presented with the idea that it is the only possible model. Potentially there may be others, but we are hard put to find any extant examples which provide better, or as good, alternative models of citizen empowerment in matters political.

There is a quite definite geographical pattern to the occurrence of the traditionally free, democratic polity. It is beyond the scope of this discussion to attempt to unravel the complex conditions which have given rise to this pattern. There is a geography to freedom; and cultural, political, and economic geographers, drawing upon history and political science, may some day assist in explicating its causes.

In concluding, let it be noted that the term "Free World" is very much a fraud when it is used propagandistically and loosely and applied by implication to all or even most of the non-communist world. The only respect in which all of the non-marxist world is free is that it is open to capitalist trade and investment without the interference of a leninist party-state. If political freedom, operationally defined, is the birthright of all human beings, then it is not a condition that should be limited to government officials whatever their ideological persuasion, or to a private entrepreneurial class, or to any other narrow segment of the people, but to all citizens of a state. We have yet a long way to go before it can be said that most of the world's people are free.

References and further reading

Anderson, T. D. 1980. *Civil and political liberties in the world: a geographical analysis*. Paper read at the East Lakes Division meeting of the Association of American Geographers, November 8, 1980.

Degenhardt, H. W. 1983. *Political dissent: a Keesing's reference publication*. Detroit: Gale Research.

Delury, G. E. 1983. *World encyclopedia of political systems and parties*. New York: Facts on File Publications.

Freedom at issue. Jan./Feb. 1986. New York: Freedom House.

Humana, C. 1983. *World human rights guide*. New York: Pica Press.

Staar, R. F. 1984. *Yearbook on International Communist affairs*. Stanford: Hoover Institution Press.

UN yearbook on human rights 1973–74 1977. New York: United Nations.

12 *Civil and political liberties in the world: a geographical analysis*

THOMAS D. ANDERSON

The amount of freedom enjoyed by individuals varies widely among the countries of the world. This complexity of conditions has many causes, including ideological differences concerning just what constitutes freedom. Disagreement exists even among scholars with similar viewpoints over the problems of how to classify and to measure levels of freedom. Such impediments notwithstanding, the subject has a broad popular interest and is a common basis for distinguishing one government from another. In order to provide a rational basis for such perceptions I have devised a system that allows all sovereign states to be ranked on the basis of the degree of personal liberty present in each in mid-1986. In a multi-authored volume devoted to the Third World this effort adds yet another perspective and set of comparisons.

The personal liberties treated here are part of a broader concept of human rights. The initial world standard in this regard was the Universal Declaration of Human Rights passed by the UN General Assembly on December 10, 1948. This document of 30 articles and 30 sub-articles is comprehensive, yet constituted merely a statement of worthy principles. Toward the purpose of establishing a legal basis for compliance by all ratifying states it was divided and passed by the General Assembly in the form of treaty provisions in 1976. These two documents are the *International covenant on civil and political rights* and the *International covenant on economic, social, and cultural rights* (US Department of State 1978).

Controversy regarding issues of human rights centers mainly on the relative significance of the content of these two covenants. The emphasis in First World (Western) countries is on the importance of civil and political liberties, whereas Second World (Eastern Europe) and many Third World governments profess a primary concern with economic and social goals. Philosophically the roots are respectively with the concept of the natural rights of man versus that of a naturally harmonious society, which is the intellectual foundation of marxism. In short, should government primarily protect individual or collective rights? A recent critique of these two positions from the perspective of East European scholars is provided by Drygalski & Kwasniewski (1986).

Domestic critics of the Western stance, however, often employ sophisms such as "other cultures have different views of freedom" or "hungry people don't care about civil and political rights." These are ethnocentric arguments that my own geographical research finds unpersuasive, especially when valid examples are not cited. There is, for example, hardly any evidence that a social justice that does not include individual justice is adopted willingly by most of the world's peoples.

The emphasis here is on the word "willingly." Fundamental to the Western concept of civil and political rights is the notion of *choice*. Choice in turn encompasses individual decisions expressed periodically in collective fashion (contested elections), in a diffusion of government power (checks and balances), and in the mutability of government policy in response to popular will (referenda). Inextricably bound up with the notion of choice are open expression of ideas and unfettered movement. In my view, denial of any of these elements constitutes infringement of basic rights, regardless of the excuses offered.

Disputes about the central content of human rights can convey the impression that the two segments are antagonistic in purpose. Yet on the assumption that social justice includes the non-discriminatory provision of such needs as food, shelter, employment, education, and old-age security, one finds that it has been achieved in countries that also protect civil and political liberties. Successful examples include (but are not limited to) New Zealand, Japan, and the Scandinavian countries. On the other hand, governments that proclaim a principal concern with collective rights do no better in social and economic areas and very much worse with respect to civil and political liberties. Examples here are Poland, Cuba, East Germany, North Korea, Libya, and Singapore.

The democratic revolution

The classification scheme used here has evolved gradually from one first devised in 1976. The focus is on civil and political liberties, but criteria more concise than those of the UN are employed. It is based on the concept of the Democratic Revolution as articulated by Preston E. James (1964, 29–31). James in turn took the term and many ideas from the work of Palmer (1959). Identified are six elements, each of which represents radical changes in the status of the individual relative to the power of the state. Oppression by rulers is an ancient condition of mankind: only its form and justifications have changed in modern times. On the other hand, legally-protected personal rights on a mass scale are little over two centuries old.

James's six elements are as follows:

(a) the individual is accorded equal treatment before the law;

(b) the individual is protected from arbitrary acts of those in authority;
(c) the individual has the right to be represented where taxes are levied or law formulated;
(d) the principle of majority rule and the use of the secret ballot are accepted;
(e) the rights of free access to knowledge and open discussion of policy issues are accepted;
(f) the individual can exercise freedom of choice: the right to take a job or leave it, to move from one place to another, to express religious convictions in any way wished – he or she possesses a revolutionary new right, the right to resign (James 1974, p. 2).

This emphasis on personal rights is not meant to belittle the importance of economic and social needs; their provision is essential to the maintenance of a society. Rather the liberating values of the Democratic Revolution raise the quality of human existence above that of a well-run military unit or slave plantation, indeed, that of an animal farm.

As a response to those who may protest that these are Western European ideas imposed on other cultures, I offer the words of Chief Joseph of the Nez Perce American Indian nation over a century ago:

> Let me be a free man – free to travel, free to stop, free to work, free to trade where I choose, free to choose my own teachers, free to follow the religion of my fathers, free to think and talk and act for myself – and I will obey every law or submit to the penalty. (As printed in the *North American Review*, April 1879.)

Although the elegant phrasing reflects his own genius, the non-literate Chief Joseph merely expressed the traditional values of his people. It seems clear that the idea that an individual has worth and a right to choose was not a uniquely European concept.

The following indicators were used to apply James's six elements. They derive in part from personal perceptions and in part from Lipset (1963, p. 27):

(a) Did the current national leadership gain power by legal means and are there one or more recognized sets of leaders attempting to gain office by the same political process?
(b) Is there an accepted legal procedure by means of which current national leaders may be removed from office or their replacements selected upon death or resignation?
(c) Do the public media openly distribute views from domestic and foreign sources that are at variance with official policy; and is news of national

events, favorable and unfavorable, freely available inside and outside the country?

(d) Are the inhabitants allowed to move freely within the country and to emigrate and return if they choose? (A related aspect is the level of restrictions placed upon foreigners who wish to leave or enter the country.)

(e) Do the country's courts make rulings against the national government, and are such rulings respected?

(f) Are there present a number of organizations not under direct state control with which inhabitants may openly affiliate if they choose?

The thrust of these indicators is to assess the opportunities for the peaceful transfer of political power, the expression of alternative views, and freedom of movement. These rights were deemed prerequisite to the function of other human rights. Restriction of movement, for example, is what most distinguishes the life of a convict from those not imprisoned. In a related sense, if reporters are not permitted to seek out and publicize details about economic, civil, political, and social conditions, how else is a society to learn about them? Surely government pronouncements are not reliable sources of news. Few North Americans accept at face value all claims by their own public officials. Why then should they not reserve judgment on unverified versions of reality by those in power in other countries?

Based on the elements of the Democratic Revolution and the supplementary indicators listed above, six categories were devised for the classification of all countries. They are:

I Countries where all elements of individual rights are specified by law and currently are extended to all inhabitants without restriction.

II Countries where all elements of individual rights are specified by law but are not extended uniformly to some minorities, often due to residual prejudice.

III Countries where most of the elements of individual rights are specified by law, but where access by many inhabitants to one or more rights is inhibited by law, custom, or arbitrary authority.

IV Countries where most of the elements of individual rights are restricted by law, custom, or arbitrary authority, but where at least one such element is available to nearly all inhabitants.

V Countries where none of the elements of individual rights is available by law, custom, or arbitrary authority, but where effective political organization provides social and economic stability.

VI Countries where the status of most inhabitants with respect to all individual rights is insecure even where specified by law, due to the capricious exercise of absolute authority or a near absence of civil organization resulting from disruptive political, social, or economic conditions.

The categories include two distinctive features. One is the attention to *all* inhabitants of a state. Many countries contain large numbers of resident aliens. To exclude such people from the *political* process is a morally and legally defensible practice. But discriminatory treatment in areas of civil liberties was viewed as an infringement of human rights.

A concern for discrimination in ethnically and racially diverse countries also is a part in that the rôle of tradition was considered. The term "custom" is used to identify cultural inertia where its presence inhibits the rights of some segments of a society. As has been demonstrated in the United States, Canada, and India, passage of legislation that mandates equal treatment regardless of ethnicity does not necessarily alter established attitudes.

The basis of country rankings

The ranking of each country was a difficult task. Verified information was used to the greatest extent possible, yet often the final choice rested on intuitive judgment tempered by long experience. Known conditions were balanced and trends considered. In several instances countries with comparable circumstances were placed in separate categories based on perceptions of progress or retrogression. An unavoidable shortcoming is the fact that bias-free information for many areas is incomplete or conflicting, and it is contemporary. Even in countries for which data are accurate, a sudden change in policy may alter human rights circumstances.

Information came from a diversity of sources, with newspapers a vital means of monitoring change. *The world factbook* issued regularly by the Central Intelligence Agency is a solid resource for background data. Comparable but different rating systems also were gleaned. These were Humana's *World human rights guide* (1984), the US Department of State's *Country reports on human rights practices* (1985), and *Freedom in the world: political rights and civil liberties, 1985–86.* The latter is prepared annually by Raymond D. Gastil for Freedom House and is of exceptional value for this purpose. My own ratings (Table 12.1) conform closely with those of Gastil despite differences in criteria and categories. Agreement was not total.

Brief analysis of country ratings

Space permits only selective explanation and analysis. In Scandinavia, for example, Iceland and Denmark are ranked higher than Norway and Sweden, not due to superior virtue but because of greater ethnic homogeneity. They

Table 12.1 Ranking of countries by civil and political liberties.

I	II	III	IV	V	VI
Austria	*Antigua-Barbuda*★	*Bolivia*	*Algeria*	Albania	Afghanistan
Barbados	*Argentina*	*Brazil*	*Bahrain*★	*Angola*	Lebanon
Costa Rica★	Australia	Comoro Islands★	*Bangladesh*	*Benin*	
Denmark	*Bahamas*	*China (Taiwan)*	*Bhutan*★	Bulgaria	
Iceland	Belgium	*Cyprus (Turk)*	*Burkina Faso*	*Burma*	
Ireland	*Belize*	*Ecuador*	*Cape Verde Islands*	*Burundi*	
Luxembourg	*Botswana*★	*El Salvador*	*Chile*	*Cambodia*	
New Zealand★	Canada	*Egypt*	*Djibuti*	*Cameroon*	
Switzerland★	*Colombia*★	Fiji	*Ghana*★	*Central African Republic*	
	Cyprus (Greek)	Finland	*Guinea*★	*Chad*	
	Dominica	*Guatemala*	*Guinea–Bissau*★	*China*★	
	Dominican Republic	*Guyana*	*Haiti*	*Congo*	
	France	*Honduras*	Hungary	Cuba	
	Germany, West	*India*	*Indonesia*	Czechoslovakia	
	Greece	Israel	*Iran*	*Equatorial Guinea*	
	Grenada	*Korea, South*	*Ivory Coast*	*Gabon*★	
	Italy	Kuwait	*Jordan*	Germany, East	
	Jamaica	*Malaysia*	*Kenya*	*Ethiopia*	
	Japan	*Maldives*	*Lesotho*	*Iraq*	
	Kiribati	*Malta*	*Liberia*	*Korea, North*	
	Mauritius	*Mexico*	*Madagascar*	*Laos*	
	Nauru	*Nepal*	*Malawi*★	*Libya*★	
	Netherlands★	*Panama*	*Mozambique*	*Mali*	
	Norway	*Peru*	*Nicaragua*	*Mauritania*	
	Papua–New Guinea	*Philippines*	*Nigeria*	*Mongolia*	
	Portugal	*Senegal*	*Pakistan*	*Niger*	
	St. Kitts–Nevis	*Sierra Leone*	*Paraguay*	*Oman*★	
	St. Lucia	*Singapore*	Poland	Romania	

St. Vincent	*Sri Lanka*	*Qatar*	*Rwanda*
Solomon Islands	*Sudan*	*Saudi Arabia*	*Sao Tome & Principe*
Spain	*Suriname*	*South Africa*	*Seychelles*★
Sweden	*Thailand*	*Sudan*	*Somalia*
Trinidad–Tobago	*Tonga*	*Swaziland*	*Syria*
Tuvalu	*Tunisia*★	*Tanzania*	*Togo*★
UK	*Turkey*	*Transkei*	USSR
USA	*Vanatu*	*Uganda*	*Vietnam*
Uruguay	*Western Samoa*	UAE	*Yemen, South*
Venezuela		*Yemen, North*	*Zaire*
		Yugoslavia	
		Zambia	
		Zimbabwe	

★ Indicates that placement in adjacent category was considered.
(Ratings were for conditions as of January, 1988.)
Third and Fourth World countries are shown in italics.

have no native Sammi (Lapps). Finland's still lower rank reflects its geopolitical decision to allow Soviet sensibilities to inhibit a full range of political expression. This sort of self-censorship is hardly repression but is not full freedom either.

Most of those countries often termed collectively "the Free World" were placed in Category II in order to highlight imperfections. In nearly all these states widespread civil and political liberties are not shared equally by various racial or ethnic minorities. The unassimilated elements consist of either indigenous peoples or recent immigrants, or both.

At the low end of the scale the human condition is as much uncertain as deplorable. Past despots such as Idi Amin and Pol Pot are out of power and instability provoked by the Soviet invasion of Afghanistan continues. The near anarchy of civil strife in Lebanon seems endless. Personal restraints in countries ranked in Category V may be greater than in those in Category VI, but they are predictable and are administered efficiently.

Category IV includes a number of obvious dictatorships and feudal monarchies. However at least one leavening freedom was distinguished in each. In many the key freedom was the right to emigrate and to return. It is the feature that most differentiates Yugoslavia from other communist-ruled East European countries, for example. Apartheid in South Africa is a despicable policy that restricts all races in some way. Nevertheless there is a steady flow of people across its borders in both directions and many opposition figures are both free from prison and quoted in the media.

All communist-governed countries are ranked in the lower half, a decision based on performance. Each practices the leninist principle of democratic centralism which makes impermissible the espousal of alternative views. The resultant repression effectively rejects all tenets of the Democratic Revolution. On the other hand, Hungary, Poland, and Yugoslavia are ranked higher not because they are less socialist but because they permit greater personal freedom.

Several generalizations seem appropriate. Of the 165 countries, 84 are ranked in the upper half. Most of these are in Western Europe, the Americas, and the Southwest Pacific. As a continent the peoples of Africa are the least free and those of North America the most free. Indeed, only a few repressive governments were in power in the entire Western Hemisphere in mid-1986. Thirty-one of the countries ranked in the upper half had historic links with the British Empire, a total that does not include Israel, Egypt, or the United Kingdom. With no apology for past British imperialism, clearly its impact included more than just merchants, missionaries, and mischief. Similarly the democracies in West Germany and Japan resulted directly from the policies of Western occupation forces after World War II.

The ranks of various countries does call into question the assumption that functioning democracy demands a highly literate population and an advanced economy. Even though most of the freest societies meet such criteria, many others do not. Striking exceptions are places like Botswana, Papua–New Guinea, and the Solomon Islands. Others with somewhat more advanced circumstances are Barbados, Belize, Costa Rica, India, and Venezuela. This evidence suggests that the premise that mass access to civil and political freedoms is a feature only of Europeanized, middle-latitude countries does not accord with reality. (Many writers express this attitude but few so clearly as MacPherson 1977, especially pp. 6–7.)

This incomplete and personalized version of the patterns of personal freedom in the world is intended as much to provoke thought as to provide information. Because it rates all sovereign states on the same scale it highlights diversity within the Third World as well as contrasts between Third World and more advanced countries. Inevitably political events will sooner or later invalidate some of the assessments shown here. Yet if this effort fosters greater awareness of the vital human issues involved it will have achieved its main purpose.

References

CIA (Central Intelligence Agency) 1983. *The world factbook*. Washington, DC: US Government Printing Office.

Drygalski, J. & J. Kwasniewski 1986. Harmonious society versus conflict-ridden society: marxism and liberalism. In *A critique of marxist and non-marxist thought*, A. Jain & A. Matejko (eds.), 259–81. New York: Praeger Publishers.

Gastil, R. D. 1986. *Freedom in the world: political rights and civil liberties 1985–86*. New York: Greenwood Press.

Humana, C. 1984. *World human rights guide*. New York: Pica Press.

James, P. E. 1974. *One world divided*, 2nd edn. Lexington, Mass.: Xerox College Publishing.

James, P. E. 1964. *One world divided*. New York: Blaisdell Publishing Company.

Lipset, S. 1963. *Political man: the social basis of politics*. Garden City, NY: Anchor Books.

MacPherson, C. B. 1977. *The life and times of liberal democracy*. New York: Oxford University Press.

Palmer, R. R. 1959. *The challenge*. Vol. I: *The age of the democratic revolution*. Princeton, NJ: Princeton University Press.

US Department of State 1978. *Human rights*. Selected documents, N. 5 (revised). Washington, DC: US Government Printing Office. (This source contains copies of eight selected international documents related to human rights.)

US Department of State 1985. *Country reports on human rights practices*. Washington, DC: US Government Printing Office.

13 *Trends in world socioeconomic development*

ALFONSO GONZALEZ

Recently there has been increased concern regarding the ability of the underdeveloped countries to maintain, let alone improve, their levels of living. Much of this is in response to the general world economy, notably influenced by the marked petroleum oil price increases of 1973 and 1979 and the subsequent inflation, balance of payments, external and internal debt, and economic recession problems that have affected most countries. These problems appear more aggravated in the underdeveloped countries, in part because of their lower levels of living and the more limited resources they have to combat these issues. There has been increased concern that economic stagnation or depression has seriously eroded the levels of living of many areas of the underdeveloped world and that this could lead to increased sociopolitical instability.

Socioeconomic development during the 1960s and 1970s

Improvement in living levels is generally quite closely associated with economic conditions. Generally economic growth was faster during the 1960s than in the 1970s except in Eastern Europe, the Orient, and, of course, the Middle East. In the 1970s beginning in 1974 (after the first major petroleum price increase), economic growth was relatively poor, especially in the industrialized democracies, subSaharan Africa, and some other major countries in other underdeveloped regions. In real terms, there was significant, although slow, improvement in the components of socioeconomic development for most of the major countries of the world. However, there were numerous countries in which absolute declines occurred.

Four measures of development – per capita GNP, diet, health, and education – were analyzed with regard to changes during the 1960s and 1970s.

PER CAPITA GNP (INCOME)

In both the industrial and underdeveloped worlds slower economic growth prevailed during the 1970s and early 1980s than in the 1960s. When compared to the industrialized countries, the greatest weakness of the underdeveloped world over the past two decades has been income growth. On a relative basis (percentage growth per capita), the underdeveloped world has been holding its own, but in *absolute* measures the gap has been clearly widening. In constant-value income, the advanced countries grew considerably faster during the 1960s than the underdeveloped countries. In the following decade the absolute growth differential was less. Of the major underdeveloped countries included in the study (Table 13.1), only Saudi Arabia during the 1970s was able to keep up with (and in fact considerably ahead of) the average of the industrial countries in growth in per capita GNP. By the early 1980s the average income of the underdeveloped countries was less than one-fifth that of the industrialized nations (it was less than one-seventh in 1960).

During the 1960s there may have been a very slight decline in real income, as expressed by per capita GNP, in perhaps only three countries (Cuba, Sudan, and Ghana). Between 1970 and 1980 real income probably declined in ten major countries (all underdeveloped): three were in Latin America (Puerto Rico, Argentina, Peru), three in the Orient (Vietnam, Bangladesh, China), and four in subSaharan Africa (Ghana, Uganda, Zaire, Ethiopia). During both decades the largest absolute income increases occurred in the industrial democracies as well as in the Middle East and Taiwan during the 1970s.

In terms of 1980 dollars, the major industrial countries increased their per capita GNP by more than $2400 in the 1960s and more than $2100 in the 1970s, while the major underdeveloped countries increased by more than $400 and $500 respectively. The gap between the rich and poor countries continued to widen, although some of the oil exporters can now be considered as the nouveaux riches.

The underdeveloped world is characterized not only by significantly smaller incomes than the industrialized world, but by a more uneven distribution of that income (see the discussion in Ch. 4). Although this may permit the accumulation of capital among the élite class, and with luck contribute to greater investment and greater economic growth, it provides an even lower level of living for the bottom sectors of the population than the national average would indicate. Many underdeveloped nations compensate for this by subsidizing basic food staples, social services, transportation, and perhaps other amenities. A more egalitarian distribution of income appears to be a concomitant of development and the more uneven distribution of income in the underdeveloped countries has little effect on country rankings (see Ch. 6).

There is only a partial correlation between income improvement (i.e., per capita GNP growth) and the rate of population growth. The industrial countries with rapid increases of per capita GNP (2.5% or greater annual increase during the 1970s) all have slow-growing populations by world standards. Some have stationary or slightly declining populations (West and

Table 13.1 Socioeconomic development indexes (1980, 1970, and 1960).

Country	Total			Per capita GNP			Diet			Health			Education		
	1980	1970	1960	1980	1970	1960	1980	1970	1960	1980	1970	1960	1980	1970	1960
USA	87	92	91	77	100	89	92	93	94	79	76	81	100	100	99
Canada	79	82	76	67	78	63	85	88	88	83	78	80	83	86	73
Australia	75	73	77	65	59	57	83	90	90	81	78	91	69	67	69
New Zealand	72	75	79	47	57	60	95	98	99	75	80	86	71	67	72
USSR	61	66	66	35	38	24	86	89	90	56	67	72	69	72	78
UK	70	68	75	57	48	59	81	87	91	78	76	86	63	63	64
W. Germany	76	69	73	73	62	74	84	83	85	78	73	73	68	60	62
France	77	76	73	69	65	62	88	90	92	84	81	79	69	68	61
Italy	68	64	61	40	37	35	91	90	81	73	68	67	68	62	60
Netherlands	77	74	79	64	51	59	82	85	85	90	92	97	73	70	73
Belgium	76	73	73	63	57	64	95	93	90	76	79	75	68	64	63
Spain	66	59	54	32	21	21	84	77	78	83	82	64	63	57	51
Sweden	83	84	85	83	85	91	80	79	85	99	99	100	73	71	62
Switzerland	85	75	85	100	70	100	85	86	92	93	87	89	62	59	60
E. Germany	70	72	66	43	52	45	92	88	84	74	77	69	72	71	64
Poland	61	62	59	25	29	21	90	90	91	63	65	61	65	64	62
Czechoslovakia	62	67	69	33	47	39	87	89	90	67	70	83	62	61	65
Romania	54	54	50	15	20	9	85	80	80	58	58	56	57	58	55
Yugoslavia	56	53	52	16	14	12	88	87	87	57	58	52	62	55	58
Mexico	46	43	40	13	14	10	67	67	65	49	51	48	53	38	38
Cuba	51	49	44	9	11	13	65	66	61	68	67	56	62	52	46
Puerto Rico	55	58	57	20	35	22			72	66	67	67	79	72	69
Venezuela	50	44	41	24	21	19	64	59	57	52	54	49	60	43	41
Colombia	40	33	35	9	7	7	55	50	54	47	48	45	51	28	35
Peru	39	36	36	8	9	8	52	57	60	41	42	37	54	35	39
Chile	49	47	45	13	15	15	67	68	67	59	50	46	56	55	51
Argentina	56	57	61	15	24	21	91	93	96	55	54	58	64	56	68
Brazil	41	39	47	13	9	8	58	61	62	46	47	45	48	39	34

Morocco	32	30	27	5	5	5	64	63	59	41	39	37	18	11	8
Algeria	33	27	26	14	6	9	57	47	51	40	40	37	21	14	7
Egypt	38	35	34	4	4	4	71	65	70	41	40	38	35	29	24
Sudan	29	25	24	3	3	4	60	57	52	34	37	32	17	5	7
Iraq	36	31	28	18	7	8	62	57	54	41	41	38	22	18	14
Saudi Arabia	56	26	26	94	9	13	71	52	56	39	34	34	18	9	2
Israel	64	64	61	34	41	26	83	84	82	72	72	75	66	60	61
Turkey	40	39	37	8	7	7	74	74	77	44	43	39	36	33	25
Iran	38	31	26	12	8	4	70	56	49	39	38	39	29	22	10
India	27	28	27	2	2	2	46	49	53	36	37	35	25	23	18
Pakistan	27	26	24	2	2	2	55	56	50	36	38	35	14	9	11
Bangladesh	23	23	27	1	2	2	41	47	47	33	28	31	16	15	
Burma	33	31	30	1	2	2	55	55	50	39	38	36	37	31	33
Thailand	38	35	34	5	4	3	50	52	49	46	47	44	53	38	39
Vietnam	35	29	30	1	3	3	47	53	52	47	34	35	46	27	
Malaysia	39	34	35	11	8	7	59	57	54	53	47	48	33	24	29
Indonesia	30	29	26	3	2	2	50	45	46	36	36	34	33	33	21
Philippines	41	40	39	5	4	4	52	48	46	46	46	44	61	62	61
China	37	34	29	2	3	2	60	54	52	52	51	33	35	28	
Taiwan	52	51	48	14	8	5		71	66	84	64	71	59	59	
S. Korea	49	43	38	11	5	4	71	64	54	53	50	48	59	53	46
Japan	76	69	60	59	40	27	75	74	69	100	91	75	70	70	69
Nigeria	28	25	24	5	3	4	53	53	54	35	30	31	18	13	8
Ghana	25	26	27	2	7	7	45	53	48	36	33	37	16	13	14
Zaire	26	23	23	1	2	2	41	46	41	34	33	33	28	11	16
Ethiopia	21	23	24	1	2	1	47	58	61	29	28	30	8	3	3
Kenya	29	29	28	2	3	3	51	60	65	40	37	34	24	14	10
Tanzania	31	23	22	2	2	2	46	48	45	37	33	34	40	9	5
Uganda	28	26	25	1	3	3	45	55	51	39	37	34	27	11	13
S. Africa	40	38	42	16	16	14	68	70	75	44	39	43	32	28	38

Sources: Essentially data from references at the end of the chapter.

East Germany); while others were, in respect of population, among the fastest-growing industrial countries (Japan and Canada). The two major industrial countries with the slowest growth of per capita GNP were New Zealand (the fastest-growing in population) and Switzerland (one of the very slowest in population growth). Among the major underdeveloped countries the relationship of per capita GNP growth and population increase is not much different. A number of countries of rapidly expanding per capita GNP exhibit slow population growth (certainly by underdeveloped world standards), e.g., South Korea, Taiwan, and Cuba; while the other countries of rapidly expanding incomes have had at least moderate population growth rates. Of the 11 major underdeveloped countries in which per capita GNP increased by about 1% or less annually during the 1970s, only two (Chile and Argentina) had slow population growth, while all the others were growing in population at relatively rapid rates.

Although the correlation between rates of economic expansion and population growth cannot really be demonstrated, the burden of rapid population growth certainly appears to have had an effect on underdeveloped countries. Of the 37 major underdeveloped countries, only four have restricted or very limited family planning services available (Argentina, Saudi Arabia, Iran, and Burma), yet in 25 countries the official policy is to reduce population growth.

The correlation of per capita GNP growth with the proportion of GDP devoted to capital investment is significantly better. The major countries that have expanded rapidly have generally devoted at least a fifth to about a third of GDP to capital investment. However, the slowly expanding economies (except in the industrial countries) have generally invested less than one-fifth of GDP to capital investment. Therefore, although population growth is a factor in the rate of economic growth per capita of countries, economic policies appear to be more significant.

DIET

The underdeveloped world generally consumes approximately one-third less calories and protein than the industrialized countries. Although the general rate of dietary improvement during both decades is somewhat better in the underdeveloped world, the industrial world has actually increased its protein intake by slightly more than, and its caloric consumption by almost the same amount as, the underdeveloped world.

Some improvement overall in the diets of the world's major countries occurred during both decades, with the greatest improvement in the Middle East during the 1970s. Undoubtedly, this was due to the infusion of oil wealth and the investment of some of that into the improvement of food supplies and greater imports of foodstuffs (the Middle East is the largest net importer of agricultural products in the underdeveloped world). This significant increase in income has also resulted in notable improvements also in health and education.

However, there was some deterioriation of diets in a number of countries. The slight declines that occurred in about half a dozen major industrial countries are not noteworthy. But in nearly a dozen major underdeveloped countries declines occurred in per capita caloric or protein consumption, or both, in both decades. Some declines occurred in some of the major countries of Latin America (Peru, Colombia, Brazil); the Orient (India, Bangladesh, Vietnam); and especially subSaharan Africa, where dietary declines occurred in Kenya, South Africa, and Ethiopia in the 1960s, and in most of the major countries in the 1970s (except the oil-exporters Nigeria and Tanzania). This is a crucial development because subSaharan Africa is already the poorest region with the slowest-growing economy, the poorest diet and the slowest-expanding agriculture, but the fastest-growing population.

Many countries of the world subsidize agriculture and food commodities in one form or another. In the underdeveloped world this almost certainly contributes to dietary improvement among the large mass of the poor. However, many countries have encountered serious economic problems during the decade of the 1970s and especially in the early years of the 1980s. In order for the countries to obtain additional foreign loans or to renegotiate payments for existing loans, a frequent prerequisite has been the reduction of government spending and the termination or reduction of commodity subsidies. This bodes ill for the mass of the population that relies on subsidized foodstuffs in order to obtain an adequate or even marginal diet.

HEALTH

In both health and education, performance has been much better than it has been with regard to diet. In both infant mortality and life expectancy the absolute improvement in the underdeveloped world has been notably better than in the industrialized countries. This is partly due, of course, to the very low infant mortality and long life expectancy that prevails in the latter countries: improvement in these elements in the advanced countries is now slow and diminishing.

Among the major industrial countries in the field of health, a noteworthy point is the significant increase in infant mortality that has occurred recently in the USSR. There has also been some increase in Bangladesh and Turkey in the 1970s and in subSaharan Africa during the 1960s. Life expectancy declined only marginally in very few countries of the underdeveloped world in either decade (Bangladesh, Ghana, and South Africa during the 1960s, and Sudan in the 1970s).

Despite significant improvements over the past two decades in the underdeveloped world, by the early 1980s infant mortality rates were still more than five times greater than in the industrial world and life expectancy was only approximately four-fifths that of the advanced countries. The Middle East and, during the 1970s, subSaharan Africa have been able to improve health conditions but both continue to have the poorest health levels among world regions.

EDUCATION

There has also been considerable attention given to education. The underdeveloped world in particular has shown considerable improvement: the rates of improvement in both literacy and enrollments in higher education have increased much faster than they have in the industrial world. Literacy levels in the advanced world are already very high, so further improvements can now be at best only very marginal. Literacy levels in the underdeveloped world are still two-thirds those of the advanced regions and the greatest improvements have occurred where literacy levels remain the lowest, in the Middle East and subSaharan Africa.

The proportion of the population in higher education has been increasing in all the major countries except for Pakistan and Kenya during the 1970s. However, here again the absolute increases have been greater in the major industrial countries. In the underdeveloped world the greatest improvements (and the highest levels) occurred in Latin America; and although subSaharan Africa has had recent improvements in literacy levels, its expansion of higher education has been only minimal.

The growth rate of higher education slowed somewhat in both the industrial and underdeveloped worlds during the 1970s. By the end of that decade the proportion in higher education in the advanced countries was still more than three times greater than in the underdeveloped world.

In general there is among the underdeveloped countries a reasonably good correlation between a high level of health or education and relatively large national expenditures in those fields. Conversely, the countries that rank low in health or education are those that have among the lowest public expenditures. The most significant exceptions to this general pattern are the oil-revenue countries, notably Venezuela and those of the Middle East, which have been investing heavily in the fields of health and education recently but which still rank relatively low in these fields. The greatest relative expenditures in education have been occurring in the Middle East and Latin America, and in health in Latin America and, to a lesser degree, the Middle East. Both the Orient and subSaharan Africa have much lower per capita expenditures in both fields.

The mean per capita GNP at the beginning of the 1980s in the major industrialized countries was nearly six times greater than that of the major underdeveloped countries, yet the public expenditures of the industrialized countries on health were more than 21 times greater and in education eight times greater than that of the underdeveloped countries. With greater incomes the industrial nations are able to spend not only greater amounts on health and education but even greater *proportions* of their incomes on such measures. Public expenditures on education are generally greater than on health, especially among underdeveloped countries. In 1979, only Bangladesh of the major underdeveloped countries spent more on health than on education.

One area in which the major underdeveloped countries (or underdeveloped countries in general) exceed the industrialized countries, considering the income disparity between the two groups, is in military expenditures. Although the underdeveloped world accounted for only 22% of the world's military expenditures in 1979, the major industrialized countries exceeded the major underdeveloped countries in military expenditures by somewhat more than three times.[1] The percentage that the mean of the major underdeveloped countries comprises of the major industrial countries in important social characteristics is as follows:

per capita GNP	17%
health expenditures	5%
education expenditures	13%
military expenditures	30%

In the major industrial countries military expenditures comprise nearly 70% of the public expenditures on education whereas in the major underdeveloped countries military expenditures exceed the combined total of both education and health by one-quarter. The only major industrialized countries in which military expenditures exceed the combined expenditures of both health and education are Israel and the USSR. In the underdeveloped world, this pattern is common in the Middle East, the Orient, and East Africa. Apparently it is in the area of military affairs that the underdeveloped world is making the greatest effort relative to the industrialized countries. During periods of economic stress the International Monetary Fund and other international agencies and banks apparently do not hesitate to recommend reduction or elimination of food subsidies and wage increases, yet they do not insist (at least not publicly) on a reduction of arms expenditures.

The socioeconomic development index

One problem is how to measure and evaluate the level of socioeconomic development and the changes that are occurring in the development process. The index that I have been using relies on four basic factors (income, diet, health, and education) which I feel are fundamental in development or in the level of living (see Ch. 4). Income is based on per capita GNP; diet is a composite of kilocaloric and protein per capita consumption; health is represented by a composite of the infant mortality rate and life expectancy at birth; and education is a composite of literacy and the proportion of the population enrolled in higher education. In each of the above elements the highest-ranking major country is considered to be 100, all the other countries being rated proportionately to the leader in that element. The overall index is the mean of the four basic factors, and each of the four are given equal weight in the index.

The major countries of the world were selected in terms of their population and GNP. A total of 58 countries were chosen (Table 13.1), slightly more than one-third of the world's sovereign countries (although one non-sovereign country, Puerto Rico, is included among the major countries). These major countries account for slightly more than nine-tenths of the world's total population and GNP.

CHANGES IN THE SOCIOECONOMIC DEVELOPMENT INDEX

During the 1960s there were 15 countries (of 58 in the study) that experienced a decline in their overall socioeconomic development index; this was reduced to 12 countries in the 1970s. However, during the earlier period, 45 of the major world countries sustained a decline in at least one factor (income, diet, health, or education) of the socioeconomic development index, while in the 1970s 39 countries did so. Therefore, most countries lost some ground in at least one important aspect of socioeconomic development relative to the world leaders. In the 1960s the overall losses were somewhat greater in Northwestern Europe, Australia–New Zealand, Argentina, Bangladesh, and subSaharan Africa (especially South Africa), while the greatest improvements occurred in Japan, Anglo-America, Spain, most of Eastern Europe, Cuba, Iran, South Korea, and China. During the 1970s the declines were greatest in the communist industrial bloc, Anglo-America, and Puerto Rico. The greatest improvements then occurred in parts of Western Europe, Japan, most of the Middle East, parts of the Orient, and Tanzania.

During the two decades only a half-dozen countries (Italy, Japan, Venezuela, Chile, Morocco, and South Korea) experienced no decrease in any of the factors of development. In per capita GNP the greatest expansion occurred in the USSR and Anglo-America during the 1960s, while relative losses were largest in Western Europe and the Middle East. However, the situation was virtually reversed during the 1970s when the greatest improvements occurred in Western Europe and the Middle East, two of the world's regions with among the best performances in economic expansion during the 1970s. Relative declines in income were most widespread among the industrial communist countries and most of the underdeveloped world outside the Middle East.

Of the four utilized in this study, diet was the only factor in which, during the two decades, the index declined in more countries than it increased in.

The Middle East again shows significant improvement in the 1970s – the only underdeveloped world region to do so. SubSaharan Africa, the poorest-fed region, exhibited declines in seven of the eight major countries, and there were important declines also in South America, South and Southeast Asia.

Health and especially education are the two fields, as indicated previously, in which world performance has been relatively good. In health the greatest improvements in the index occurred in subSaharan Africa and most of the underdeveloped world, while the industrial communist countries sustained the most significant declines. In education there was a significant improvement during the 1970s compared to the 1960s, and during the decade of the 1970s only five countries (Canada, the USSR, Romania, the Philippines, and Taiwan) showed a decrease in the index. The greatest improvements were in all the underdeveloped regions, particularly during the 1970s.

COMPONENTS OF THE SOCIOECONOMIC DEVELOPMENT INDEX[2]

There is considerable variation in almost all countries when one compares the individual components of the index. In most countries at least two (frequently three) components vary at least 10 points from the overall index. Taking into account notably that per capita GNP depresses, and diet inflates, the overall index, there are more than a dozen countries which vary relatively little among the four factors when compared to the total index. These are New Zealand, Italy, East Germany, Czechoslovakia, Mexico, Venezuela, Brazil, India, Bangladesh, Burma, Malaysia, Indonesia, Ghana, and Zaire.

Most countries, however, rank significantly higher or lower in one or more aspects of development than in the other components.

PER CAPITA GNP

Considering their overall level of socioeconomic development, the following countries rank much higher than expected: many of the industrialized democracies, and some of the petroleum-rich Middle Eastern countries, notably Saudi Arabia and Iraq. However, Eastern Europe, Cuba, Puerto Rico, Argentina, the Philippines, and underdeveloped East Asia rank much *below* what would be expected.

DIET

The countries that rank relatively high are Argentina and much of the Middle East, whereas Japan is quite low.

HEALTH

The relatively high-ranking countries are Spain, Sweden, Netherlands, Japan, Cuba, China, and Taiwan; while the USA, the USSR, Argentina, and Saudi Arabia are very low.

EDUCATION

The very high-ranking countries are the USA, much of Latin America, especially Puerto Rico, and the Philippines; while the relatively low-ranking countries are Switzerland, Sweden, Pakistan, and much of the Middle East and subSaharan Africa.

On a regional basis, Western Europe, Anglo-America and, recently, the Middle East rank relatively rather high (considering the other components of socioeconomic development) in income whereas Eastern Europe, the USSR, and Latin America have relatively depressed incomes. However, the situation is practically reversed with regard to diet. In this factor it is Western Europe and Anglo-America that rank little above their level of development (however, they are still the best-fed regions), while the Middle East, Eastern Europe, and the USSR have a dietary level significantly higher than would be expected from their overall level of development.

In health, regional variations are not very great, and to a lesser degree the same is true of education, although in the former the USSR and Anglo-America rank somewhat low and the Orient rather high. In education, Latin America is somewhat advanced and the Middle East held back by the factor of education. A tendency of note is that there is a very slow narrowing of the differences in the development factors herein studied, with the notable exception of income (per capita GNP) in the underdeveloped world. The differential in the latter has been slowly increasing, indicating that although the underdeveloped countries may be having some success in closing the gap with the wealthier leaders in terms of diet, health, and education, few countries appear to be closing the income gap.

Summary and conclusions

The pace of socioeconomic development has varied considerably over the past two decades both with regard to time and among individual countries and world regions. In general, the industrialized countries have slowed their rate of socioeconomic improvement in the 1970s and early 1980s, compared with the 1960s, in all the measures utilized in this study. The underdeveloped countries overall show a mixed performance in the 1970s compared with the 1960s. However, the underdeveloped countries have expanded relatively faster than the industrialized nations, especially during periods of recent slower growth. Only in the Middle East, the Orient, and Eastern Europe did the overall economies of the major countries expand faster in the 1970s than in the 1960s, whereas the rate of population growth has declined in all regions, except the Middle East and subSaharan Africa. The net result is that GNP growth per capita among the major countries was faster in the 1970s in the Communist bloc and the underdeveloped world, except notably in subSaharan Africa.

The major underdeveloped countries have generally been improving their level of socioeconomic development relatively faster than the industrial nations. However, they are at a considerably lower level to begin with and the absolute improvement of the underdeveloped countries has generally been less. The early 1980s saw a period of serious worldwide economic recession and certainly proved a setback for most countries in socioeconomic development. If 1983–4 proves to have been the turnaround period, economic and social conditions may be seen to have improved later in the 1980s.

The Middle East exhibited the greatest relative improvement during the 1970s and this was undoubtedly due in large measure to the inflow of petroleum revenues. Half of the major Middle Eastern countries are great petroleum exporters and these have suffered also in the early 1980s. Furthermore, two outstanding performers during the 1970s, Iran and Iraq, have been engaged in a war since September 1980. The pace of development in the Middle East will depend strongly on the international petroleum market and on the political stability of the region.

Two problem regions that have shown disappointing performance, as measured by this socioeconomic development index, during the 1970s are the communist industrial nations and subSaharan Africa. The latter region had an even poorer performance in the 1960s. Although the rate of overall economic growth in the communist bloc during the 1960s was rather impressive, it has not translated into comparable improvements in consumer goods or in my measure of socioeconomic development. SubSaharan Africa ranks lowest among world regions in virtually all measures of socioeconomic attainment. Of the underdeveloped regions it is improving overall the least; and in one component, diet, it has deteriorated.

In attempting to close the gap between the industrial and underdeveloped worlds (at least with regard to the measures used in this study), there are some interesting and relevant points to keep in mind. During recent periods of economic recession the slowing of economic growth has been greater among industrialized economies. Furthermore, there is a limit to how much further the industrial nations can improve their diets – there are already indications that many are beginning to reduce caloric and protein intake. A somewhat similar situation exists with regard to health. Infant mortality is now very low in the industrialized world and little quantitative improvement is possible in the future, while life expectancy has been increasing very slowly among the leaders. Again the underdeveloped countries can reduce the gap significantly since improvement among the leaders will probably not be noteworthy until there is a major breakthrough in geriatrics when life expectancy may again be increased significantly. In literacy, industrial countries are almost all very close to the 100% level, but considerable expansion is very possible in higher education. In the latter aspect the underdeveloped world will have more trouble closing the gap.

Latin America is clearly the most advanced of the underdeveloped regions, while the Middle East has improved fastest during the 1970s. If the trends of the 1970s are projected into the future, the Middle East overall would attain the level of Latin America during the second decade of the next century. Should the industrial communist bloc, the least advanced of the industrial world, halt its relative decline and remain basically unchanged into the future, the Middle East would attain the level of the bottom of the industrial world in the third decade of the next century.

Should the trends of the 1970s continue, the mean of the underdeveloped countries will not attain the level of the industrialized nations for another century and half. However, there is some variation among the underdeveloped regions, most notably the relatively sluggish performance of the poorest region (subSaharan Africa), and considerable variation among the countries within the regions. In any case, it will take a long time for the development gap overall to narrow considerably, especially if the industrial world resumes its rapid growth and if the underdeveloped world cannot expand faster than it has in the past.

Notes

1 Based on data from Ruth Leger Sivard: *World military and social expenditures 1982*, Tables II and III.

2 The four components that comprise the overall total index each has equal weight. However, the components have different effects on the overall index (see Ch. 4). Per capita GNP generally depresses the index – most countries have incomes much below the leader in comparison to other components and there appears to be little limit as to how much income can be attained. Diet is just the opposite: it inflates the overall index – there is a definite upper limit to how much humans can consume, there is also generally a lower limit, and the range is proportionately not very large. Nevertheless, both the health and education components approximate the overall index most closely, and much more closely than do diet or per capita GNP. Both consist of two elements each, and one in health (infant mortality) and one in education (literacy) have definite upper limits. If one were to rely for an indicator on only *one* of the factors of socioeconomic development used in this index, it would appear that education or health would be the best choice for most countries since these components generally represent levels closest to the overall index.

References

FAO production yearbook 1981 (and earlier editions). Rome: Food & Agriculture Organization.

Sivard, R. L. 1982 (and earlier editions). *World military and social expenditures*. Leesburg, Va.: World Priorities.

UNESCO *statistical yearbook.* 1982 (and earlier editions). Paris: United Nations Educational, Scientific & Cultural Organization.

World Bank atlas 1983 (and earlier editions). Washington: World Bank.

World development report 1983 (and earlier editions). Washington: World Bank.

World population data sheet 1983 (and earlier editions). Washington: Population Reference Bureau.

World population estimates 1981 (and earlier years). Washington: The Environmental Fund.

PART III

Major Third World Regions

Our thoughts bring us to diverse callings, setting people apart: the carpenter seeks what is broken, the physician a fracture, and the Brahmin priest seeks one who presses Soma.
O drop of Soma, flow for Indra.
With his dried twigs, with feathers of large birds, and with stones, the smith seeks all his days a man with gold.
O drop of Soma, flow for Indra.
I am a poet; my Dad's a physician and Mum a miller with grinding-stones. With diverse thoughts we all strive for wealth, going after it like cattle.
O drop of Soma, flow for Indra.
The harnessed horse longs for a light cart; seducers long for a woman's smile; the penis for the two hairy lips, and the frog for water.
O drop of Soma, flow for Indra.

The Rig Veda (c. 1200–900 BC)
Trans. by Wendy D. O'Flaherty
(1981, Penguin Books)

This last part studies the status and problems of development in almost all of the major regions of the Third World. In addition, a special perspective is provided on the Third World circumstances occurring in industrialized countries.

Thomas Anderson describes the subregions of the Caribbean Basin and the diverse and contrasting political conditions in one of the least militarized regions (except for marxist regimes). The region ranks fairly high in development within the Third World and there are also significant resource differences among the countries. It is a region of emigration, reliance on foreign trade, and sociopolitical tensions, with the greater range of options residing in the larger mainland countries.

Alfonso Gonzalez, like most of the other authors in this section, probes the unifying and contrasting physical and cultural characteristics of South America. There are significant national differences, and economic development has resulted again in both a variety of common traits and significant diversity among the countries. The problems of development are numerous; the levels of socioeconomic accomplishment are several; and the range is impressive.

Raja Kamal and **Hal Fisher** describe the great diversity of the Arab World and divide the countries into three economic groups. Major problems appear to be the Arab–Israeli conflict, the Iranian impact, the oil-glut, and inadequate planning. Social change is occurring and there are three major social groups vying for influence, along with divergent political philosophies.

B. L. Sukhwal also probes the elements for a considerable degree of unity, despite great diversity, in South Asia. Some historical background is also provided. Development has apparently made a considerable advance since independence. Planning is in general a major aspect of development, with the region showing varying degrees of success. The biggest problem appears to be the population explosion. Cultural diversity typifies the region as does a broad range of political systems.

Thomas Anderson indicates the common social traits, the material cultural elements, and the physical diversity of East Asia, the Chinese culture realm. Clear national identities occur in this region, which is little associated with colonialism. The diversity of the continent (inner) and marine (outer) countries is pointed out. China and North Korea, despite communist governments, he considers as Third World. An analysis is provided of recent historical events in this region of basically two divided countries (China and Korea). The contrast in development strategies between the marxist (central planning) and non-marxist (export-oriented) governments is quite marked. Family-planning programs have reduced fertility to the lowest levels in the Third World and these have been accompanied by social and economic changes. The author feels that success may be due to adaptability, discipline, industriousness, and education, all combined with political stability.

Hal Fisher analyzes the numerous major development problems of subSaharan Africa. Regression has occurred over the past decade and internal and external programs have failed to resolve fundamental problems. There appear to be two approaches, modernization with reliance on international markets, and disengagement with greater national or regional self-reliance. The author feels that more comprehensive policies are necessary to promote greater determination, self-reliance and regional cooperation.

The final two chapters provide a very different perspective on development. The approach of Evelyn Peters and Joel Lieske is to study two industrialized countries and analyze a less developed subculture within those countries.

Evelyn Peters presents a case study of the Canadian Indians, an indigenous group who are part of what some call the Fourth World: Evelyn Peters calls this the Third World within a First World. The circumstances of

Canadian Indians parallel those of many underdeveloped nations, but aboriginal peoples are differentiated from people of the Third World. The Indians are demographically and socioeconomically distinct from other Canadians. Their cultural traits face challenges from the dominant society and they have little control over changes. This has been disruptive of their societies and the author's prescription is for increased Indian self-determination and self-government.

Joel Lieske analyzes 243 US metropolitan areas using 123 indicators for the overall quality of life. He finds a racial dualism with a growing divergence between black and white incomes, uneven development between areas and regions, and cultural/political pluralism. Quality of life is lower in metropolitan areas with high concentrations of blacks (but not Hispanics), and the author feels that the social costs of dualism are felt by everyone. He also advocates that priority should be for human (over purely economic) development; and that superior quality of life (at least in the US) is not due to a diversified economy or manufacturing productivity but to the development of an educated middle class, and so is most enhanced by a moralistic (participant) rather than an individualistic (commercial) or especially a traditionalistic (parochial) political culture.

A. GONZALEZ

14 *The socioeconomic "worlds" of the Caribbean Basin*

THOMAS D. ANDERSON

The Americas consist of two broad socioeconomic components. The United States and Canada occupy most of the North American continent; have cultural, economic, and political similarities; and are First World countries. The rest of the hemisphere is termed "Latin America" and is composed of developing countries (Fig. 14.1). This chapter focuses on the Caribbean Basin, the most diverse part of the Americas. Addressed briefly are differing national features, attainments, and potentials in the context of common regional concerns.

The Caribbean Sea is the region's integrating element. It provides a maritime routeway between the two main landmasses, as well as between the Atlantic and Pacific Oceans by way of the Panama Canal and land routes across the narrow isthmus of North America. The region is adjacent to the world's leading superpower, the United States, which historically has feared hostile power influence there. This proximity and the associated attitude inescapably affect US relations with Caribbean countries.

With 25 countries and nine overseas dependencies, the Caribbean Basin is the world's most politically diverse region. All the countries in it except Haiti and Cuba are part of the Third World. Haiti's socioeconomic indicators are sufficiently below those of its neighbors to place it in the Fourth World. Yet despite this lowest ranking in the hemisphere, Haiti's per capita GNP in 1981 was higher than those of 17 countries in Asia and Africa (see Table 14.1).

Cuba is classed here as part of the Second World, an assessment made with full knowledge that Fidel Castro from 1979–83 was head of the Non-Aligned Nations organization. However, its Second World credentials include marxist–leninist forms of government and economy, membership in the Council of Mutual Economic Assistance (CMEA), which features full economic integration with Second World countries, and close military coordination with the USSR.

Sub-regions of the Caribbean Basin

Augelli (1962) has identified "Rimland" and "Mainland" segments of the Caribbean, based on differences in historical development. The Rimland

Figure 14.1 Political units of Latin America.

consists of the islands and coastal fringes of Central America and the Guianas. Here European conquest destroyed native peoples and the subsequent labor needs of plantations caused a re-population first by slaves from Africa and, following emancipation, by contract workers from South Asia. Modern populations reflect these African and East Indian origins and the culture residue of slavery and plantations permeates Rimland societies. The Rimland's centuries-old reliance on international commerce persists despite recent changes in basic activities and trade partners.

Despite population declines due to conquest, most native peoples on the Mainland were acculturated under Spanish rule. The contemporary Mainland populations consist racially of various proportions of Amerindians, whites, and mestizos. The hacienda, not the plantation, was the most influential institution. Its economic function was mainly to serve domestic needs rather than foreign markets. Its legacies include economic and class inequalities and a rural-based power structure that long dominated national affairs.

Such generalizations are not intended to obscure the uniqueness of each state or the spread of modernization. Their purpose is to assist intra-regional comparisons.

A distinction between Spanish and non-Spanish entities also is pertinent. All countries of the Mainland are Spanish, except Belize, as are Cuba, Puerto Rico, and the Dominican Republic. All share the broader Latin American culture. Independence came earlier in the Spanish lands, the most recent being Panama from Colombia and Cuba from Spain around 1900.

Haiti's independence from France dates from 1804. However, in other non-Spanish lands independence began only in 1962 with Jamaica and Trinidad and Tobago beginning a decolonization process that continues. Languages and cultures here reflect colonial affiliations. Hence Dutch is spoken in Suriname and the Netherlands Antilles; French in French Guiana, Guadeloupe, and Martinique; and English everywhere else.

This process has created a cluster of increasingly smaller states of questionable economic and political viability. Just how many remaining dependencies will choose independence remains unclear. The French holdings are Overseas Departments (DOMS), politically equivalent to *départements* in France. The relation of the Netherlands Antilles to the metropole is similar. British dependencies are Crown colonies except for the Associated State of Anguilla. As a Commonwealth, Puerto Rico has the choice of independence, statehood, or continuation of the status quo. Supporters of the last two options in the past have claimed over 90% of the vote.

Aside from Suriname the new Caribbean mini-states are English-speaking, a factor that diversifies the interests of the region and has geopolitical consequences. During the British–Argentine conflict over the Falkland/Malvinas Islands, for example, voting in the Organization of American States (OAS) consistently split along Spanish/English lines.

Democracy and militarism

A notable feature of the Caribbean Basin is the large number of governments chosen by popular vote in contested elections. Under such democratic rule the inhabitants have the rights to speak, write, and move about freely, as

Table 14.1 Selected data on Caribbean Basin countries.

	Total population*	Annual natural increase* (%)	Infant mortality rate*	Age 15+/64+* (%)	Life expectancy at birth* (years)	Literacy† (%)	Urban* (%)	Per capita GNP* 1982 $ USA	Per capita GNP rank in 1981§	Civil–political liberty rating‡	Armed forces per 1,000 population¶	Total debt: 1983** ($ millions)
North America												
Belize	0.2	2.5	27	44/4	70	90	52	1 130	75	1–1	–	58
Costa Rica	2.7	2.6	19	35/3	73	93	48	1 070	64	1–1	1.5	2 669
El Salvador	5.1	2.4	67	45/3	65	57	39	680	91	2–4	5.4	911
Guatemala	8.6	3.1	71	45/3	59	51	39	1 110	73	4–4	2.3	1 309
Honduras	4.6	3.2	82	48/3	60	47	37	670	96	2–3	3.9	1 383
Mexico	81.7	2.6	53	42/4	66	83	70	2 180	51	4–4	2.0	60 386
Nicaragua	3.3	3.4	76	48/3	60	87	53	880	81	5–5	27.8	2 363‡‡
Panama	2.2	2.1	26	39/4	71	85	49	2 110	56	6–3	5.0	2 821
South America												
Colombia	30.0	2.1	53	37/4	64	82	67	1 410	66	2–3	2.6	7 229
Guyana	0.8	2.2	35	37/4	68	96	32	560	90	5–5	8.8	672
Suriname	0.4	2.0	31	40/4	69	80	66	3 390	43	6–6	5.0	15
Venezuela	17.8	2.7	39	40/3	69	86	76	3 830	36	1–2	3.2	16 538
Caribbean Islands												
Antigua–Barbuda	0.1	1.0	21	28/6	72	*89*	34	1 690	–	2–3	–	–
Bahamas	0.2	1.9	22	38/4	69	*90*	69	4 050	40	2–2	–	176
Barbados	0.3	0.9	11	29/9	73	*98*	73	3 990	42	1–2	3.3	229
Cuba	10.2	1.1	15	28/8	73	96	70	1 461††	–	6–6	23.5	5 103‡‡
Dominica	0.1	1.7	13	41/6	74	*94*	74	970	89	2–2	–	–
Dominican Rep.	6.4	2.5	64	41/3	63	62	52	1 160	68	1–3	4.1	1 780
Grenada	0.1	1.8	22	36/6	71	*98*	–	830	83	2–3	–	–
Guadeloupe	0.3	1.4	14	31/8	72	–	46	–	35	3–2	–	68
Haiti	5.9	2.3	108	39/6	53	21	53	290	127	5–5	1.4	398

Jamaica	2.3	1.8	21	37/6	73	82	54	1 270	70	2–3	0.9	1 773
Martinique	0.3	1.1	14	28/8	73	–	71	4 260	–	3–2	–	48
Neth. Antilles	0.3	1.3	26	33/5	71	–	66	–	–	2–1	–	623
Puerto Rico	3.3	1.4	16	32/8	74	91	67	3 800	–	2–1	–	–
St. Kitts–Nevis	0.04	1.8	43	37/9	64	–	45	1 320	–	1–1	–	–
St. Lucia	0.1	2.5	18	50/5	71	*82*	40	1 050	77	1–2	–	–
St. Vincent	0.1	2.5	37	44/5	65	*96*	–	840	94	2–2	–	–
Trinidad–Tobago	1.2	1.8	24	34/6	70	97	23	6 830	28	1–2	1.7	1 127

Sources
* Population Reference Bureau 1986.
† James & Minkel 1986, *passim*. Numbers in italics, *82*, adapted from Population Bureau, 1982.
‡ Gastil 1986, pp. 7–13. Numbers are political rights – civil liberties rated 1–7, with "1" the freest conditions.
§ The World Bank 1984.
¶ US Arms Control and Disarmament Agency 1984, pp. 16–52.
** OECD 1983, pp. 91–212. (‡‡ does not include debt to Soviet Bloc countries.)
†† Wilkie & Perkal 1984, p. vii.

well as to organize unions and political opposition. Nonetheless repressive governments of diverse ideology were present in 1987. The regimes in Cuba, Guyana, and Nicaragua were avowedly marxist. Even though no right-wing dictatorships remained, authoritarian rightist governments ruled in Haiti and Suriname. Freedoms also were restricted in various ways in El Salvador, Guatemala, Mexico, and Panama. The military establishment held at least latent power in each of these states except in Mexico and Guyana. Improvements in the direction of greater freedoms began early in 1986 in Guatemala and (one hopes) Haiti.

Caribbean Basin countries are among the least militarized in the world. Costa Rica, Belize, and the new island states have only token forces. Defense of the dependencies is provided by the metropolitan powers. Data compiled in 1982 showed military expenditures of less than 2% of the GNP in 11 of 17 Caribbean countries. These were Barbados, Colombia, Costa Rica, Dominican Republic, Haiti, Jamaica, Mexico, Panama, Suriname, and Trinidad and Tobago. The greatest expenditures were in Cuba, Honduras, and Nicaragua, which spent 5–10% (US Arms Control and Disarmament Agency 1984, p. 4). In terms of number of armed troops per 1,000 population Jamaica ranked lowest with 0.9 and Nicaragua highest with 27.8 (see Table 14.1).

Marxist regimes were the most militarized. Below Nicaragua's 27.8 per 1,000 population were Cuba with 23.5 and Guyana at 8.8, whereas the next highest was El Salvador at 5.4. Cuba's regular forces of 225,000 in a population of 10 million contrasted sharply with Mexico's 95,000 troops in a population of 78 million. By early 1985 active military forces in Nicaragua had increased to 62,000 (US Department of State 1985, p. 1). This number was greater than the combined total of the other six Central American states, nearly equal to those of Colombia, and half again larger than those of Venezuela. These latter states are much more populous. (For more on this issue see Payne 1985, p. 16.) These exceptions notwithstanding, relatively small proportions of Caribbean budgets are devoted to militarization.

Economic Development

Compared with the Third World as a whole, Caribbean Basin countries ranked well on the basis of per capita GNP in 1981 (see Table 14.1). Of 145 national economies evaluated by the World Bank, Caribbean countries held from 28th to 127th place. First World and oil-exporting countries held the top 36 positions, but of the next 62 national economies, 21 were Caribbean. Haiti was the lone low-ranked exception. Space constraints allow only summarization of the region's socioeconomic circumstances. Most countries are grouped according to collective criteria, as much to illustrate certain themes as to inform. Several, however, merit individual treatment.

Mexico and Venezuela are the most developed and may be regarded as

Newly Industrialized Countries (NICs). They are home to roughly half the region's people. Both are major petroleum exporters with reserves expected to last at least several decades longer. Proximity to the United States is a marketing advantage but oil price declines in the 1980s have severely reduced national revenues. Each nationalized its petroleum industry and the income flows directly to the government. Both have invested heavily in social programs and the national infrastructure, yet within each mismanagement and corruption have diverted major sums from their planned purposes. Venezuela is a founding member of OPEC, whereas Mexico does not belong. Following the large oil price increases in 1973 they jointly subsidized oil sales to oil-poor neighbors in the Caribbean, an action that simultaneously increased their influence and lessened that of the United States.

Heavy borrowing on the expectation of continued high oil demand left each country heavily in debt to First World banks with the onset of a worldwide petroleum glut in the 1980s. The amounts change with time but in 1983 Mexico owed some $60 billion and Venezuela about $17 billion (see Table 14.1). Such massive debt accumulation is perhaps better understood by recognition that it occurred during the general development surge 1960–80 during which Latin American economies as a whole tripled in size (Enders & Mattione 1984, pp. 2–7). Both domestic and international factors have contributed to the regional debt crisis exemplified by Mexico and Venezuela.

In addition to oil and gas Mexico exports a wide diversity of minerals. Venezuela is a major exporter of iron ore, and potentially of bauxite, but its other minerals at best supply merely national needs. Multinational investments have spurred diversified industrial growth in both countries, a growth enhanced in part by mandatory domestic component regulations. Tariff-shielded manufacturing in Venezuela mainly serves the national market, whereas Mexican industrial products are increasingly world-market oriented.

Colombia stands on the threshold of NIC status. Its assets include large size and population, strategic location, and varied mineral and agricultural resources. Its coal deposits are many times larger than any others in Latin America and in the mid-1980s large coal exports began from Guajira Peninsula fields. Discovery of new petroleum reserves increased this known resource from sufficiency to exportable proportions. Unlike Mexico and Venezuela, Colombia's main export was agricultural – coffee – and its foreign debt of over $7 billion, although a burden, was not so intimidating.

Colombia's leading rôle in the hemisphere's illegal drug trade, however, was both an economic boon and a sociopolitical curse. Massive amounts of marijuana are grown and shipped in the northeast and Colombia also is the site of cocaine processing from coca leaves grown mainly in Peru and Bolivia. The enormous sums of money that pass into the hands of criminal elements profoundly disrupt the country's social and political fabric. The trade also has caused diplomatic tensions with the United States, the main target of the drug flow.

In the countries of Central America conditions differ considerably. Neither minerals nor fossil fuels are produced in quantity, and aside from oil deposits in Guatemala prospects for greater development are slight. Foreign exchange in Central America is earned mainly from the exports of coffee, bananas, sugar, and cotton (in descending order of value) except in Panama. Panama's top-ranked per capita GNP is derived largely from Canal tolls and from commerce, manufacturing, and tourism stimulated by the presence of the Canal. It is also one of Latin America's leading financial centers. Light manufacturing in Central America enjoyed about two decades of expansion until a broad downturn in the 1980s.

The sag in the economies reflected a concurrent worldwide recession but there were local causes as well. Late in the 1970s leftist rebels in El Salvador and Guatemala began campaigns of sabotage against the national infrastructures. By 1985 rebellion in Guatemala had been subdued but conflict continued in El Salvador. Nicaragua's economy after 1970s was harmed by a sequence of events that included a major earthquake, a destructive civil war, systematic looting of financial assets by the ousted Somaza clan and, after 1979, mismanagement by the new leftist government accompanied by attacks on economic targets by anti-Sandinista guerrillas.

More basic than such disruptions is the fact that Central American countries produce nothing to sell that is not also available from other Caribbean Basin sources. Its human resources also are not outstanding in quality. Although literacy and health levels are quite good in Belize, Costa Rica, and Panama, they are at best only moderate in the other four countries. Efforts to increase tourism faced stiff competition from Mexico and the Caribbean islands.

The Caribbean islands overall also has few minerals. The main exceptions are Trinidad and Tobago which has substantial petroleum for export and Jamaica which is a major bauxite producer, Bauxite reserves also are important on the mainland in Guyana and Suriname. Exports of Cuba's large deposits of nickel and iron ore suffered from a competitive world market. Only minor amounts of copper, nickel, and bauxite are mined in Haiti and the Dominican Republic. Elsewhere minerals are virtually absent.

Historically the wealth of the islands came from exports of tropical crops and of sugar and rum in particular. For over a century Cuba has been a leading world sugar exporter. Yet in 1987 it was the only place in the Caribbean still mainly dependent upon its traditional crops of sugar cane and tobacco. Jamaica's most valuable export also is a crop – marijuana – which is neither traditional nor legal. Elsewhere on the islands sugar cane, bananas, coffee, cacao, citrus, or spices are of varying importance but their former share of the national economies has declined. Throughout the islands, other than in Cuba, plantation production has decreased and the number of small commercial farms has increased.

On most islands tourism exceeds agriculture as a source of income. It is the paramount industry in the Bahamas, Antigua-Barbuda, Barbados, St. Martin, the Cayman Islands, and several even smaller islands. Tourism shares significance with petroleum processing in the Netherlands Antilles, US Virgin Islands, and St. Lucia, and is a major component of the more complex economies of Jamaica, Haiti, Dominican Republic, and Trinidad and Tobago. Important causes of expanded tourism were the US trade embargo of Cuba and the development of jumbo-jet aircraft. Negative aspects of tourism are its seasonality, low wages, proportionately large foreign ownership, and unpredictable changes in tourist preferences. These elements, plus growing popular dissatisfaction with the servile nature of most jobs, have caused island political leaders to seek alternative income sources (Anderson 1984, pp. 67–71).

Manufacturing expansion has focused on labor-intensive industries integrated with First World economies. Regional advantages are literate workers, underemployment, low wages, ocean transportation, tax rebates, and close historical links with the First World. The dependencies benefit also from intra-national overseas markets, with the success of the "Bootstrap" growth in Puerto Rico after 1950 a notable example. Cuba's membership of the CMEA is a socialist variation of such an economic relationship. Haiti and the Dominican Republic, where the international trade affiliations are weakest, not surprisingly also have the lowest socioeconomic level. In these respects they compare with most states in Central America.

The geopolitical perception of increased Soviet influence in the region motivated the United States in 1983 to implement the Caribbean Basin Initiative. Its purpose was to stimulate economic expansion through the mechanisms of private investment and tariff reductions. Initial funding was modest. Innovative features of the initiative were its treatment of a multi-country region as a unit and the involvement of Canada, Mexico, and Venezuela as co-sponsors. Its eventual impact is uncertain but it can be regarded as a positive approach to a regional problem.

Migrations and geopolitics

Mass human movements have characterized the Caribbean Basin since the Columbian discoveries. Early destruction of the island natives led to repopulation not just by slaves from Africa but by lesser numbers of Europeans and Asians as well. Over the past century a restless flow of island workers have sought seasonal and temporary jobs both within the region and in Anglo-America and Europe. The largest numbers of these people came from Haiti, Puerto Rico, and Cuba, with many of the latter motivated by political circumstances since 1960. Mainland migrants from Mexico and Central America to the United States now total up to 10 million, most of whom have arrived since 1945. Since 1960 Venezuela attracted perhaps 2 million migrants, most from Colombia.

These migrants are not always welcomed, nor do all enter legally. They are no less real, however, and they affect societies at both ends of the migration stream. Workers overseas collectively send or bring back to their Third World homelands billions of dollars. Such money reaches needy families without a bureaucratic filter and provides maximum purchasing benefits. Through personal foreign experiences or by means of correspondence ordinary people even in remote communities become well informed regarding First World conditions. This awareness introduces new standards and sparks dissatisfactions, especially among youth.

In the United States, Canada, France and the United Kingdom, clusters of Caribbean Basin residents add ethnic diversity and also exert growing political influence on domestic and international issues. These First World countries are becoming more conscious of their Caribbean linkages. Indeed mass migrations have caused Miami, Florida, to be regarded not only as a Caribbean city but even as its regional capital. Miami's loosely-fettered, continual interchanges of people, capital, goods, and ideas, along with its polyglot population, make it an urban interface of the First and Third Worlds.

The Cuban Revolution since 1960 has affected the entire region. Fidel Castro's efforts to bring radical change to neighboring countries and Cuba's integration into the Second World brought an actual rather than a potential East–West dimension to Caribbean affairs. Traditional US concern about the security of its southern border and vital sea routes was heightened by Cuba's military alliance with the United States' most feared rival, the Soviet Union. US reactions have included armed intervention in the Dominican Republic in 1965 and Grenada in 1983, as well as efforts to overturn the leftist government of Nicaragua. One profound impact has been the trade embargo against Cuba in 1961. The action caused Cuba to seek trade partners elsewhere and not only in the Soviet Bloc. It also increased the share of the lucrative US market for other Caribbean Basin countries. The tourist industries in particular benefited.

The emergence of regional power centers has diluted United States hegemony in the region. Besides Cuba (whose activities are monitored closely in Washington), Mexico, Venezuela, Colombia, and Panama also exert influence beyond their borders. These states not only pursue their own outside interests but acting as the Contadora Group they have sought peaceful solutions to conflict in Central America. Each has opposed the United States on various issues. Mexico in particular serves the rôle of a non-aligned nation, so much so that it does not even participate in the organization of that name. As a policy Mexico avoids identification with recognized blocs aside from the Organization of American States.

Prospects for development

Although the Third World countries of the Caribbean Basin face obstacles in their efforts at greater economic development, there are regional assets as well. A basic problem is that each national economy depended on foreign trade. This dependence was magnified during the world recession of the early 1980s because of a growing protectionist mood in much of the First World, as well as the general need for food imports, a need as real in socialist Cuba as in the other island states. Venezuela also is a net food importer. Other Mainland states largely are able to feed themselves but their agricultural surpluses are of nutritionally inessential commodities such as coffee, sugar, and bananas. Wide wealth disparities and high rates of underemployment contribute to social and political tensions. Even Cuba, where income gaps are more narrow, is not free from discontent.

Such problems notwithstanding, the region's developmental assets are considerable. It has a central position in the Hemisphere and abuts the First World. Other than in Haiti literacy levels are high for the Third World and are rising. Most inhabitants live in open societies with freely-chosen governments regularly accountable to their constituents. Although not a quanitifiable development factor, an environment of individual freedom is conducive to all manner of innovations. Contrary to the popular image, widespread political stability prevails, not only in the democracies but also in dictatorships such as Cuba. Even in Central America civil strife is confined to El Salvador, Nicaragua, and to a diminishing extent Guatemala. In only a few countries are large proportions of the national budget diverted for military purposes. Despite an uneven distribution, the region as a whole has large resources of energy and minerals. The tropical climates favored the export of tropical crops to and tourism from the colder climes of the First World.

The greater resource bases and advanced economies of Mexico, Colombia, and Venezuela give them a greater range of development options, including that of a largely domestic orientation. Strategies available to the small, mineral-poor countries of Central America and the Caribbean islands are more limited. Ideologically the basic choices are between versions of a centrally-planned or a market economy. Because Cuba's rate of socioeconomic development under Castro has been less than that of its island neighbors, its socialist system has not been an attractive model for emulation (see Horowitz 1984, *passim*). More to the point, Soviet-style economies everywhere have been consistently poor in technology innovation, consumer goods production, and commerce. Yet it is precisely these three categories that offer the best prospects for development in the small states of the Caribbean Basin.

A review of national development worldwide shows that neither a paucity of resources nor small size precludes economic progress. For example, high per capita GNPs have been achieved in Switzerland, Sweden, Denmark, Iceland, Finland, the Netherlands, Austria, New Zealand, Israel, Hong Kong, Taiwan, and Singapore (Lewis & Kallab 1983, pp. 207–21). Of course, programs that succeed in one culture may not work so well in another. Nonetheless, the consistent feature is that in each of these countries creation of national wealth occurred in market systems closely integrated with the world economy. It appears that the left-of-center elected governments of the Caribbean Basin have drawn similar conclusions and have shaped national policies accordingly. So long as the region's peoples retain a political choice one can anticipate that development strategies of a similar nature will continue.

References

Anderson, T. D. 1984. *Geopolitics of the Caribbean: ministates in a wider world.* New York: Hoover Institution and Praeger.

Augelli, J. P. 1962. The rimland-mainland concept of culture areas in Middle America. *Annals of Association of American Geographers* **52**, 119–29.

Enders, T. P. & R. Mattione 1984. *Latin America: the crisis of debt and growth.* Washington, DC: The Brookings Institution.

Gastil, R. D. 1986. Comparative survey of freedom 1986. *Freedom-at-Issue* **82**, 3–17.

Horowitz, I. L. (ed.) 1984. *Cuban Communism*, 5th edn. New Brunswick, NJ: Transaction Press.

James, P. E. & C. Minkel 1986. *Latin America*, 5th edn. New York: Wiley.

Lewis, J. P. & V. Kallab (eds.) 1983. *U.S. foreign policy and the Third World: agenda 1983.* New York: Praeger.

OECD 1983. *External debt of developing countries: 1983 survey.* Paris: Organization for Economic Cooperation and Development.

Payne, J. L. 1985. Marx's heirs belie the Pacifist promise. *Wall Street Journal.* April 5, 16.

Population Reference Bureau 1982. *World's children data sheet.* Washington, DC: Population Reference Bureau.

Population Reference Bureau 1986. *World population data sheet.* Washington, DC: Population Reference Bureau.

The World Bank 1984. *World tables: the third edition.* Vol. I: *Economic data.* Baltimore: Johns Hopkins University Press.

US Arms Control and Disarmament Agency 1984. *World military expenditures and arms transfers, 1972–1982.* Washington, DC: Government Printing Office.

US Department of State 1985. *The Sandinista military build-up.* Publication 9432, InterAmerican Series 119. April. Washington, DC: Government Printing Office.

Wilkie, J. W. & A. Perkal (eds.) 1984. *Statistical abstract of Latin America.* 23, Los Angeles: UCLA Latin American Center Publications.

15 *South America: the worlds of development*

ALFONSO GONZALEZ

The continent of South America[1] comprises the southern nearly nine-tenths of the cultural region of Latin America but contains only two-thirds of that region's total population. It is a continent that exhibits a considerable diversity of physical characteristics, some of them quite extreme, as well as outstanding cultural features and contrasting levels of socioeconomic development.

Almost one-third, the southern portion, of the continent ("the southern cone"[2]) lies outside the tropics. Mid-latitude climates extend northward to include virtually all of Paraguay (certainly the effectively settled national territory), the southern part of the Brazilian state of Mato Grosso, virtually all of São Paulo, and into southern Minas Gerais.

South America possesses some truly interesting physical superlatives. It contains the highest continuous mountain barrier on earth (the Andes) with the highest summits outside of central Asia. The Andean region has a larger population residing at high altitudes than anywhere else and it is here that some of the world's highest cultural features are found. The world's greatest river, by far, in volume, drainage basin, and navigability (the Amazon) dominates the interior of the continent. The world's highest as well as the greatest waterfalls (in volume) are found in the eastern highlands, and nearly one-fifth of the world's potential hydroelectric power is on this continent. No region in the world has such a high proportion (one-half) of its area in forest, and within the same country, Chile, are located both the driest weather station and the wettest place (in number of days with precipitation). One of the direct effects of the physical environment is that the continent, like Latin America overall, is affected by natural disasters more often than any other region except the Orient.

In cultural characteristics South America also has some unique features. It, along with Middle America[3], is the only major world region that not only has a relatively large proportion but significant numbers of all three major racial stocks. It is the most complex and intermixed region racially, although race is not a major issue. It is, by far, the most Roman Catholic region on earth, and it contains nearly two-fifths of the world's Catholics. Religion, notably the status and influence of the church, has played a major rôle in Latin American affairs and it is still a factor to contend with.

South America was the first area of the underdeveloped world to attain independence, yet, like Middle America, it is the only underdeveloped region that adopted and retained European languages and religion and that contains large numbers of people of European extraction. Furthermore, in virtually all measures of development, it is the leader among underdeveloped regions.

General unifying characteristics

The most outstanding common characteristic that ties the countries of the continent together is the three-century long period of colonization that ended in the second and third decades of the 19th century. Although Brazil was under the control of Portugal and the other countries were divided among three Spanish viceroyalties, the long colonial experience was quite similar.

The colonial period resulted in the development of a number of cultural traits that distinguish virtually all of Latin America. European languages and religion were superimposed and adopted so that no other underdeveloped region has such a large area that is as relatively homogeneous in both language and religion or so westernized.

Population has been strongly concentrated in a small portion of each national territory, with outlying areas being generally neglected and developing a very limited infrastructure. All countries, except Uruguay, have a large proportion of their total area that is very sparsely settled or developed. The interior half of the continent contains no more than 10–15% of the total population.

The fundamental sociopolitical structure, with an established élite, consisting, at least in the early stages of development, of the landed oligarchy, was also established essentially during the colonial period and was strongly reinforced after independence.

The limited development of an entrepreneurial class and restricted effective investment has characterized all countries until perhaps recently. The general absence of direct US intervention, in contrast to the Caribbean Basin, has also typified the continent.

National differences

The physical environment of individual countries is rather varied. Coastal lowlands are everywhere rather restricted but three great lowland river basins dominate much of the continental interior. The very high and rugged Andean cordillera on the western margin makes transportation and communication in that region difficult. Geologically older plateaux occur on the eastern margins of the continent.

Arid conditions prevail over two large areas of the continent: (a) in the northern third of Chile and extending along the entire Peruvian littoral, and (b) in Patagonia (the southern one-quarter of Argentina) extending into the northwest of Argentina and adjacent Bolivia. Smaller subhumid conditions occur in portions of northeastern Brazil and in the western coast of Venezuela and adjacent Colombia.

Vegetation types are very varied and frequently present problems to the development of agriculture and transportation. Humid tropical soils that prevail over the lowlands of most of the continent are generally low in quality and easily damaged by improper and/or frequent cultivation. However, some of the world's best soils are located in the Pampa region of Argentina. As indicated previously, mid-latitude climates as well as vegetation and soil types predominate over about the southern third of the continent.

Resources, like all other features (physical or cultural), are unevenly distributed in South America as elsewhere. High-quality agricultural land occurs in Argentina, Uruguay, South and Southeast Brazil, and central Chile. Many minerals are most abundant in the central Andean region, Chile, the Brazilian plateau, and select smaller areas. Petroleum and natural gas are most abundant in Venezuela, with smaller deposits in Ecuador, Argentina, and some other countries. In addition, South America is a major world producing region of iron ore, copper, lead, zinc, tin, and silver.

High cultural levels were attained in the pre-Columbian period in the central and northern Andean regions. Currently, the Amerindians comprise a large proportion, even a majority, of the population from southern Colombia to Bolivia. They still resist integration into the national societies and these regions today comprise some of the least developed areas of the continent. Large numbers of Europeans (primarily from Iberia and Italy) migrated to Argentina and to São Paulo and south Brazil, with smaller numbers going to Uruguay and middle Chile. These areas are certainly the most commercialized and industrialized in South America. Negroids comprise an important component of the population, especially in Northeast Brazil and in the coastal lowlands from Ecuador to Venezuela. Asians (Japanese) are significant only in São Paulo (excluding, as was indicated, the Guianas).

There are significant differences in other population characteristics as well. Growth rates vary from the relatively slow pace in the mid-latitude countries (due essentially to lowered fertility) and the fast growth of the central Andean countries and Paraguay (due to higher fertility and somewhat higher mortality). The degree of urbanization also varies from more than 80% in those slow-growing mid-latitude countries to less than 50% in Paraguay, Ecuador, and Bolivia.

Attempts at different forms of political structure have been tried by various South American countries. Ostensibly Argentina, Brazil, and

Venezuela have federal forms of government but, in reality, they are also centralized in function. Monarchy was tried in Brazil for more than half a century after independence; Chile experimented for a time with a parliamentary system, and Uruguay with an executive council instead of the presidential system that has generally prevailed. A true revolution occurred in Bolivia (1952) and major attempts at fundamental reforms have also occurred in Peru and Chile.

Prolonged periods of democratic administrations occurred in Chile and Uruguay (both terminated in 1973), with shorter intervals in Colombia and Argentina. At the moment, and South America does go through political cycles, democracies prevail in all countries except Chile and Paraguay. Some countries have had Caudillos of relatively long duration, like Paraguay which has the longest-reigning (since 1954) head of state in the Western Hemisphere. Other countries are far more unstable, such as Bolivia which since independence in 1825 has had about 190 coups, 70 presidents, and 11 constitutions.

Some institutions that are frequently involved politically include the Church, which varies from strong in Colombia and Peru to weak in Uruguay. Labor unions have the strongest traditions in Bolivia, Argentina, Uruguay, and, except under the present regime, Chile.

Another factor to be taken into account is the continuing animosity between neighboring countries arising from territorial disputes. Hostilities have occasionally broken out and the attitudes of government have been instrumental in difficult relations between neighbors, seriously hindering attempts at regional integration. The major territorial disputes involve Venezuela/Guyana, over the Essequibo region currently within Guyana; Ecuador/Peru, over the Amazonian territory and their boundary; Bolivia/Chile, over Bolivia's loss of its outlet to the sea in the War of Pacific, 1879–83; Bolivia/Paraguay, over the Chaco; Argentina–Chile, over the Beagle channel and the islands just south of Tierra del Fuego; and Argentina/UK, over the Falkland/Malvinas Islands.

Economic development

The economies of the countries of South America have a number of outstanding characteristics in common: (a) they are in varying degrees of underdevelopment; (b) all place considerable emphasis on economic expansion; (c) most have programs and all stress the need for improvement of the poor masses; (d) all experience the relative decline of agriculture in employment, gross domestic production (GDP), and exports; and (e) all show the more rapid growth of the industrial sectors of the economy. However, there are also some important differences among the countries: (a) in the stages or levels of socioeconomic development; (b) in the rates of economic growth

Table 15.1 South America: population and economic characteristics.

	Population 1986 (m)	Population growth (p.a.) 1973–82	Urban (%)	Per capita GNP 1983 (US $)	GNP growth 1973–82 (%/year)	Population economically active (%)		Proportion of GDP 1983		Exports 1981	
						Agriculture 1983	Manufacturing c. 1980	Agriculture (%)	Manufacturing (%)	Primary (%)	Manufacturing (%)
Venezuela	17.8	3.5	76	4,100	3.5	16.1	15.0	6.6	17.9	96.1	1.9
Colombia	30.0	1.9	67	1,410	4.6	24.8	14.6	22.7	20.4	71.6	27.3
Ecuador	9.6	2.6	45	1,430	5.7	42.5	11.3	13.3	19.1	96.8	3.0
Peru	20.2	2.4	66	1,040	2.0	35.2	11.7	14.1	22.3	59.1	40.8
Bolivia	6.4	2.6	47	510	1.5	48.3	10.1	16.6	14.7	61.5	38.5
Paraguay	4.1	2.5	43	1,410	9.4	47.8	13.2	29.7	16.0	89.3	10.1
Chile	12.3	1.6	83	1,870	3.0	17.0	14.5	9.4	19.9	28.4*	71.6*
Argentina	31.2	1.3	83	2,030	0.2	12.1	21.4	15.4	23.7	75.6*	24.4*
Uruguay	3.0	0.5	85	2,490	3.4	11.0	20.7	12.2	20.9†	70.1*	30.0*
Brazil	143.3	2.3	69	1,890	5.2	35.8	18.1	8.4	24.1	58.8	39.6
South America‡	277.9	2.1	66	1,818	3.9	29.1	15.1	14.8	19.9	70.7	28.7

Sources: Based on various editions of Population Reference Bureau, *World development report*, *World Bank atlas*, FAO *Production yearbook*, UN *Statistical yearbook*, IDB *Economic and social progress in Latin America*, and ILO *Yearbook of labour statistics*.

* 1975. † Includes mining. ‡ South America: average for the countries.

and socioeconomic change; (c) in the relative importance of agriculture and industry; and (d) in the degree of government ownership and intervention in the economy.

There has been a considerable difference in the rates of growth of the South American countries (Table 15.1). Per capita growth was overall faster during the 1970s than in the previous decade. In the 1960s Brazil, Colombia, and Argentina were growing moderately fast while Uruguay and Ecuador were the slowest-expanding. During the 1970s Brazil, Paraguay, and Ecuador grew rapidly; Peru and Chile experienced periods of growth and decline; and Argentina expanded very slowly overall. The early 1980s saw a period of relative decline, especially for manufacturing.

A universal trait of development or modernization is the relative decline of agriculture in an economy. All the countries of South America experienced a decrease in the proportion of the economically active population and the GDP accounted for by agriculture. The mean for the South American countries of the population engaged in agriculture decreased from more than two-fifths in 1960 to less than 30% in 1983 (Table 15.1). In the latter year the more agricultural countries (with more than a third to nearly one-half of the population in agriculture) consisted of all the small countries (except Uruguay), Brazil, and Peru. One-quarter of Colombia's population was still in agriculture while in the southern cone countries and Venezuela the proportion was only between a ninth and a sixth of the population. Although considerable strides were taken in the expansion of the industrial labor force in virtually all countries, by the early 1980s more people were still engaged in agriculture than in manufacturing in all countries except Argentina and Uruguay. However, in Venezuela and Chile both sectors were about equal.

All the countries have long reached the stage where the urban population is growing faster than that of the rural/agricultural sector. A further stage in settlement evolution occurs when the agricultural population begins to decline in absolute numbers (as is occurring in all the advanced countries). Argentina and Uruguay entered this stage decades ago. More recently, Chile, then Colombia also, began to experience an absolute decrease in their agricultural populations. In the middle and late 1970s Venezuela also entered this stage. In the other countries the agricultural population continues to increase but at an ever-declining rate. This slowdown or reversal of the growth of the agricultural population, along with the massive rural–urban exodus, helps to explain the declining emphasis on land reform as a major issue.

Although agriculture remains pre-eminent as a form of employment, the same cannot be said of its value of production. Since the 1950s the value of manufacturing has exceeded that of agriculture on the continent. The value contributed by agriculture declined in all countries during the 1960s and 1970s, with a slight reversal during the deep recession of the early 1980s. Manufacturing contributed increasing proportions of national production in all countries (again, with a slight reversal in the early 1980s). Colombia, Bolivia, and especially Paraguay remain the only countries on the continent where agriculture continues to exceed manufacturing in value. During the 1960s and 1970s manufacturing was expanding more than 50% faster than agriculture and, obviously, was receiving priority by virtually all governments.

The export of manufactured goods has also characterized South America as countries have moved from a policy of import substitution to a more externally-oriented approach to industrialization. Only in Venezuela, Ecuador, and Paraguay did manufactures comprise less than a quarter of total exports in the early 1980s.

Agricultural products now contribute a smaller share of the export trade (about 30%) but about the same share of the imports (slightly more than 10%). Most countries have increased their cereal imports due to relatively rapid population growth, the increasing importance of higher-quality livestock production, and slightly improved diets (Peru being a significant exception), but not importantly to a shift to export crops or livestock raising.

Traditionally, the mineral-oriented economies, Venezuela, Bolivia, and Chile, have been net importers of agricultural commodities. Bolivia has reduced that dependency but Peru in the 1980s has become a net importer. All the other countries continue as net exporters and the continent remains one of the two world regions (with Anglo-America) having the greatest net agricultural surpluses. South America's net surpluses are primarily in tropical products, such as coffee, sugar, cotton, bananas, cacao, and citrus, but also in meats, mid-latitude fruits, tomatoes, and recently soybeans.

US direct investments are most noteworthy in Brazil, with lesser amounts in Argentina, Peru, and Colombia, and very little in the small countries. Tourism is most important in Brazil, Argentina, Colombia, and Uruguay, but is really not comparable with most industrialized countries or Mexico (the underdeveloped world's leader).

Government absorbs a gradually increasing share of GDP in most countries, while gross domestic investment has significantly increased only in the northern Andean countries and Paraguay, with a notable decrease in Peru and Chile during the past two decades. The degree of government participation in the economy also varies significantly. In Colombia, Ecuador, and Chile, government involvement is probably the least on the continent while public ownership and intervention is of a greater magnitude in the others, with perhaps the most in Uruguay.

Generally a problem that is not as evident in Latin America as in other regions is military expenditures. Only about 2% of South America's GNP is devoted to the military, and this is the lowest proportion anywhere. However, this apparently is no restraint on military intervention in politics where coups are more frequent than anywhere except subSaharan Africa. Despite relatively low military expenditures, these do exceed both education and health expenditures in Uruguay (it is to be hoped that this may now change), and health expenditures in all other countries except Brazil, Colombia, and Venezuela.

Some problems of development

In the process of development there are a number of persistent problems that confront these countries, but the problems typify the underdeveloped world and generally become more manageable with development.

RAPID POPULATION GROWTH

Although during the past decade the rate of population growth has begun to slow down, it still exceeds 2% annually in virtually all countries outside the southern cone and is at least 2.5% in the smaller countries (Table 15.1). Family planning programs were initiated in the late 1960s and early 1970s and fertility has dropped significantly in the tropical countries, except Bolivia where mortality is also the highest in the hemisphere. Although rapid population growth does not cause poverty or underdevelopment, it does present a formidable obstacle that an economy must overcome in order to raise the general level of living.

INADEQUATE CAPITAL INVESTMENT

Domestic capital available is insufficient and frequently conditions are not propitious to attract foreign funds. Foreign investment and assistance provide considerable controversy and are political problems that must be faced.

EXTERNAL DEBT

Brazil and Argentina have two of the largest foreign debts in the underdeveloped world, but all the other countries now have to contend with the problems of repayment or rescheduling. There are groups who place the blame (or a large part of it) on the banks of the industrialized world, but poor planning, mismanagement, and the lack of prudence on the part of the recipient underdeveloped countries certainly contributed to the problem. Although the threat of loan defaulting appears to have passed for the present, the austerity measures have resulted in a decline of real living levels and pose a constant threat of sociopolitical instabilty.

INFLATION

As elsewhere this problem is ameliorating somewhat but prior to the rapid inflation of the 1980s, the southern cone was the most inflationary area in the world and Bolivia in the mid-1980s had the greatest rate anywhere.

MONOCULTURAL EXPORT ECONOMY

The dependency on one or two commodities characterized most Latin American economies but in the past decade or so there has been a widespread endeavor to diversify.

INCOME DISTRIBUTION

The underdeveloped world is characterized by a markedly uneven distribution of income and Latin America probably has the most uneven distribution of all. Although no country of South America for which data is available is comparable with the industrialized countries, the southern cone countries have the least uneven distribution of all of Latin America. On the other hand, countries like Brazil and Peru have markedly uneven distribution by any standards. Consumer expenditures and investment potentialities are both restricted by this pattern.

Levels of socioeconomic development

The range of development that exists in South America is as great as can be found in any other underdeveloped area. There is a marked difference between the degree of development or level of living in countries such as Argentina and Bolivia. Based on the socioeconomic development index (Ch. 4), which employs four basic components (per capita GNP, diet, health, and education) the fairly broad diversity of development levels can be seen (Table 15.2).

Argentina has clearly been South America's most developed country, although its relative position in the world has been declining gradually. Uruguay which earlier was closer to Argentina has declined relatively even more and is now about on a par with rapidly-improving Venezuela and slower-progressing Chile. The next group comprises Ecuador, Paraguay, and Colombia, with Brazil and Peru farther down the scale and Bolivia at the bottom. During the past two decades the greatest overall improvements appear to have occurred in Ecuador and Venezuela, both the beneficiaries of greatly augmented petroleum income, especially during the 1970s when their improvement exceeded that of the previous decade. Colombia, Bolivia, and Chile experienced modest but significant gains, while improvement overall in Brazil and Paraguay was slight. Peru underwent a very slight decline but the relative decreases in Argentina and Uruguay are noteworthy.

Progress overall for South America seems to have been better in the 1960s than the 1970s, and this holds true for diet and education. Health improvements were generally greater in the 1970s and the relative decline in incomes (i.e., per capita GNP) was less in the 1970s.[4] Income improvement has been very difficult for South America, like all other underdeveloped regions, except for the oil-rich and newly industrialized countries.

Table 15.2 South America: comparative development.

Country	Total index			Per capita GNP			Diet			Health			Education		
	c. 1980	1970	1960	*c.* 1980	1970	1960	*c.* 1980	1970	1960	*c.* 1980	1970	1960	*c.* 1980	1970	1960
Venezuela	50	46	42	25	21	22	66	60	57	53	53	49	57	49	40
Colombia	41	38	36	9	7	8	57	52	54	47	46	45	50	45	35
Ecuador	42	36	34	9	6	7	50	51	47	46	44	41	64	45	39
Peru	37	40	37	6	9	11	55	59	60	41	44	38	45	48	38
Bolivia	32	33	27	3	4	4	51	51	44	36	35	35	39	41	27
Paraguay	42	42	41	9	5	6	72	73	68	49	46	48	37	44	42
Chile	49	46	45	11	15	17	70	70	67	59	47	46	57	52	53
Argentina	56	59	62	12	24	25	94	96	96	54	54	58	62	60	68
Uruguay	51	54	58	15	17	24	76	83	86	54	58	63	57	56	60
Brazil	39	39	38	12	9	9	60	63	62	45	46	45	39	39	34
South America*	44	43	42	11	12	13	65	66	64	48	47	47	51	48	44

Sources: Based on data of various editions of the *World development report*, *World Bank atlas*, Population Reference Bureau, FAO, *Production yearbook*, and UNESCO *Statistical yearbook*.

* South America: average for the countries.

Diet improved significantly in Venezuela and Bolivia, but has deteriorated in Peru and Brazil and less importantly also in Uruguay (which by underdeveloped world standards still has outstanding nutrition). Health improvements have been most notable in Chile and Ecuador. In education South America has shown the greatest progress and all countries except Argentina, Paraguay, and Uruguay have undergone some relative improvement.

Summary and conclusions

The continent of South America provides contrasts in both the physical landscape and cultural characteristics so that the individual countries have to contend with differing circumstances in their quest for socioeconomic development. The approaches and policies of different countries vary, as do those in the same country through time as administrations change, frequently by coups.

Countries are confronted, to varying degrees, by population growth, inadequate capital investment, debt and inflation, a relatively narrow-based economy, and a host of social and political problems and obstacles. Overall, there are different levels of development and the degree of change also is quite marked. Virtually all countries, as measured over the past two decades, are improving in real terms and on a per capita basis; but on a relative basis, compared with the world overall, the results are far more mixed. By most measures of development and modernization the countries are changing but whether the change is fast enough is in serious question.

There are some interesting economic points with reference to this continent that should not be overlooked. Perhaps the greatest nationalistic fervor regarding nationalization occurred in Argentina (railroads), Bolivia (tin), and Chile (copper). All these enterprises have now been nationalized at least for more than a decade and none is doing well, for a variety of internal and external reasons. Regional integration has been attempted through the Latin American Free Trade Association (LAFTA; 1960) and the Andean Group (1969) and despite some initial success have foundered essentially because of internal problems. Land redistribution programs have been attempted in various countries and invariably food production has declined and food scarcity has resulted. Emphasis on private enterprise, the open market, and foreign investment have been tried with notable success (in some aspects) for a time at least in Brazil and not so successfully in Chile. A policy of greater government intervention, to a degree probably unprecedented in South America, was tried in Chile with unsatisfactory results at least.

The governments of these countries, whether through the electoral system or the fiat of an élite, have some interesting questions or options to consider. Who is to develop and control the mineral resources (or other resources, for that matter)? Most surely would prefer to control the resource but do not possess sufficient capital or technology to develop it. In a somewhat similar vein, what about agribusiness, especially if foreign? There appears to be a fetish now about "dependency on imported food," but if imports are less costly than local production (food or manufactured goods) the result that must be faced is that consumer prices will be higher or government subsidies will be costly. Although other considerations may be involved (encouragement of local producers, security of production, perhaps social benefits), governments must take the economic costs into account.

The way ahead for South America is certainly not clear. The potential is considerable but the problems are formidable. The options have to be weighed and the question is whether the decision will be made via a free and honest electoral process or by an élite group (with or without popular support). And the basic question regarding policy to be resolved is who will benefit and who will bear the cost.

Notes

1 "South America" as used in this chapter refers to the whole of that continent except for the Guianas (Guyana, Suriname, and French Guiana).

2 The "southern cone" refers to the southern portion of the continent of South America. It comprises Chile, Argentina, Uruguay and the South and perhaps the Southeast of Brazil. In this study, the country of Brazil is not included as a southern cone country.

3 Middle America refers to the northern 13% of the Latin American region (i.e., the southern portion of the North American continent). Middle America comprises the mainland and islands between the USA and South America, and therefore includes Mexico, Central America, and the Antilles (West Indies).

4 Although incomes, even in real terms, increased for virtually all South American countries in both decades, compared to the world leaders (in the industrialized world) many countries experienced a *relative decline* in their income position.

References

FAO 1983 (and earlier editions). *Trade yearbook 1982*. Rome: Food and Agriculture Organization.

FAO 1984 (and earlier editions). *Production yearbook 1983*. Rome: Food and Agriculture Organization.

ILO 1984 (and earlier editions). *Yearbook of labour statistics 1984*. Geneva: International Labour Organization.

Inter-American Development Bank 1984 (and earlier reports). *Economic and social progress in Latin America*. Washington: Inter-American Development Bank.

Population Reference Bureau 1985. *World population data sheet 1985*. Washington: Population Reference Bureau.

Sivard, R. L. 1985 (and earlier editions). *World military and social expenditures 1985*. Washington: World Priorities.

UN 1985 (and earlier editions). *Demographic yearbook 1983*. New York: United Nations.
UN 1985 (and earlier editions). *Statistical yearbook 1982*. New York: United Nations.
UNESCO 1983 (and earlier editions). *Statistical yearbook 1983*. Paris: United Nations Educational, Scientific and Cultural Organization.
US Department of Commerce 1985. US direct investments abroad . . . 1984. *Survey of Current Business* **65**,8, 30–46.
World Bank 1985 (and earlier editions). *World Bank atlas 1985*. Washington: World Bank.
World Bank 1985 (and earlier editions). *World development report 1985*. New York: Oxford University Press.

16 *Change and development in the Arab world – advance amid diversity: an economic and social analysis*

RAJA KAMAL and HAL FISHER

An economic analysis

Many mistakenly think of the Arab World as a homogeneous region unified by Islam and the Arabic language. However, reality indicates the opposite: the Arab World is highly diversified. The 15 countries discussed in this chapter include the richest in the world, such as Kuwait and the United Arab Emirates (UAE), and some of the poorest (e.g., the Sudan). Some are conservative (e.g., Saudi Arabia) and some are not (e.g., Libya and Syria). Some are heavily populated (e.g., Egypt), and others (e.g., Oman) have barely a million inhabitants. Some claim a long, rich national heritage and identity (e.g., Iraq), while others (e.g., UAE and Oman) have existed as independent nations less than two decades. Thus, although the Arab nations lack homogeneity, all of them desire advance and development.

The Arab countries discussed here may be classified as "developing." The oil price escalation of 1973 brought economic enhancement to many rich Arab oil-exporting countries. One finds it difficult, if not impossible, to deal simply with economic development in the Arab World: each nation is approaching development in its own way. Usually internal conditions hinder development within the Arab World, but external causes also contribute.

After the dramatic increase of the price of oil since 1973, "Arab World" became an economic as well as a geographic expression. The flow of oil revenues from all over the world represented a turning point for the entire Arab World, but especially for the oil-producing nations, who channeled some of their income surpluses to other developing Arab countries, either as loans for development or as direct grants. The emerging rich, oil-producing nations (Saudi Arabia, Oman, UAE, Kuwait, Libya) represent a small portion of the Arab nations discussed here; the rest may be classified as lower-middle-income and upper-middle-income countries. The 15 nations fall into three economic groups:

(a) high-income oil exporters (Saudi Arabia, Libya, Kuwait, UAE, and Oman);
(b) upper-middle-income (Algeria, Iraq, Jordan, and Syria);
(c) lower-middle-income (Egypt, North and South Yemen, Morocco, Sudan, and Tunisia).

HIGH-INCOME OIL-EXPORTING COUNTRIES

Although the world prices of oil started rising in 1971, the dramatic increase took place in 1973, the year of the Arab–Israeli Yom Kippur war. The price rise provided sizeable revenues to all oil-exporting countries and particularly for the nations with large oil reserves, such as Saudi Arabia and Kuwait. The increased revenues became an effective source for fast economic development.

Today, the oil-exporting countries have gained virtually full cost control over their indigenous natural resources. Their immediate use of revenues generated has led to significant economic changes, especially in Saudi Arabia, Kuwait, and the UAE. These nations have struggled both with their own rapid development and with meeting the responsibilities of Islamic and other developing nations worldwide.

With the exception of Libya, the Arab oil-rich nations share good political and economic relations with the United States and Western Europe. Their per capita GNP (Gross National Product) ranks among the world's highest. They chose to use their oil revenues for infrastructure. For example, Saudi Arabia and Kuwait built highways and telecommunications systems which are the best in the Middle East and among the best in the world. They built hospitals, schools and universities in a short time. The oil-exporters provide excellent health services to their citizens, as the life-expectancy data in Table 16.1 reflect. Lack of blue- and white-collar workers did not slow development; laborers and professionals were imported from all over the world as they were needed.

It is difficult to ascertain the magnitude of the financial flows from the Arab oil-exporting countries to others in the region. Most private transfers and some of the public ones are not well documented or publicized; however, transfers of funds on a regular basis were never disrupted. To aid others, the principal oil-producing states have created autonomous development funds based on concrete technical and economic grounds. For example, in 1961 Kuwait established the Kuwait Fund of Arab Economic Development (KFAED), with an initial capital of 50 million Kuwaiti Dinars (KD). By 1974, the budget of KFAED had reached KD 1.0 billion and its aid assisted all Arab developing countries. Saudi Arabia and the UAE have established similar funds.

Inter-Arab labor migration forms one indirect but important source of financial aid from the oil states to the other Arab countries. Again, 1973 was the turning point, and the subsequent oil-price jump the reason. Oil revenues pouring into the Arab oil-exporting nations increased their need for

Table 16.1 The Arab world: basic indicators.

Country	Population 1983 (m)	Population growth projection–2000 (m)	Per capita GNP 1983 ($US)	Daily per capita calorie supply 1982 (% of requirement)	Life expectancy at birth 1983	
					Male	Female
Lower-middle-income						
Egypt	45.2	63	700	128	56	59
Yemen PDR	2.0	3	520	97	45	47
Yemen Arab Republic	7.6	12	550	97	43	45
Morocco	20.8	31	760	110	51	54
Sudan	20.8	33	400	96	47	49
Tunisia	6.9	10	1,290	111	60	63
Upper-middle-income						
Algeria	20.6	38	2,320	110	55	59
Iraq	14.7	26	–	118	57	61
Jordan	3.2	6	1,640	117	63	65
Syria	9.6	17	1,760	123	66	69
High-income oil exporters						
Oman	1.1	2	6,250	–	51	54
Libya	3.4	7	8,480	152	56	59
Saudi Arabia	10.4	19	12,230	129	55	58
Kuwait	1.7	3	17,880	–	69	74
UAE	1.2	2	22,870	–	68	73

Source: World Bank 1985.

imported labor. Saudi Arabia alone increased its imported laborers from 345,000 to 700,000 workers in less than five years. Saudi Arabia, Kuwait, the UAE, and Libya became the major importers of Arab labor; and Egypt, the two Yemens, and Jordan the major exporters. The inter-Arab labor migration was and still is successful for the following reasons:

(a) Within the labor-importing countries, with some exceptions, the incoming laborers have melded easily into the society they entered. Both already shared religion and language.
(b) The imported labor force working on infrastructures obtained experience which they later transferred to their home countries.
(c) The Arab labor force working and residing in the rich Arab nations transferred substantial amounts of their wages as savings and investments to their home countries, thus helping those economies.
(d) For the exporting countries, the increase of their workers laboring abroad meant a decrease of unemployment. Egypt, with the largest surplus of workers, benefited the most.

During the past decade, the Arab oil-exporting countries have faced major challenges. Not only have they had to struggle with obstacles to economic development, but they have also had to fulfill the obligations they took toward the rest of the Arab world and many Third World countries. Once seen as a center for trade and growth, the region has slowed economically and has lost some financial credibility because of these challenges. Although the adverse factors are primarily exogenous to the region, their impact is nevertheless clear and obvious. They are:

(a) The continuing Arab–Israeli conflict. For the Arab nations involved in this conflict (Egypt, Jordan, Syria, and Lebanon), the economic assistance of the richer Arab countries is vital to continued financing of the costly struggle.
(b) The Iranian impact. The fear of a possible pro-Khomeini Iranian control of the Gulf has prompted other Arab states in that region fully to support Iraq against Iran. If Iraq were to lose, the less powerful rich Gulf countries could be next in line. The philosophical differences between some Arab countries are also clearly evident in the Iran–Iraq war. Thus, the Iranian influence could change the political and economic complexion of the entire region.
(c) The oil glut worldwide. Due to depressed demand for oil, most producing countries have been forced to cut back oil production. The Organization of Petroleum Exporting Countries (OPEC), which consists of some of the world's oil-producing nations, including those in the Arab Gulf region, has lost its economic clout and perhaps its cartel image. Between 1979 and the mid-1980s, OPEC's total production dwindled to about 40% of its capacity. For the first half of 1985, the oil production of the United Kingdom (a non-OPEC member) exceeded that of Saudi Arabia, once the world's leading oil exporter. The production cutback by Saudi Arabia probably severed the last effective bond uniting OPEC. The diminishing oil revenues to the Arab World disrupted national plans and development goals set in the late 1970s and early 1980s. At the time of writing, world oil prices have dropped sharply, partly because Saudi Arabia has flooded the market with its competitively priced oil.
(d) Lack of proper planning in the early stages. In the past, income from oil revenues was used to finance massive development projects. The balances were invested in financial ventures and institutions throughout the world. Today, there is common understanding among the policy makers of the rich Gulf region that their main natural resource – oil – should be transformed into a reproduction asset dependable in both long and short runs.

UPPER-MIDDLE-INCOME COUNTRIES

Countries in the upper-middle-income classification (annual per capita GNP between US $1600 and $6000) have progressed well industrially and are now at similar economic levels in their development. Syria, Iraq and Jordan, along with Lebanon and the Palestinians, all in the same geographical region, represent much of the Arab World's administrative and technical élite. In the Arab World, their share of the agricultural and industrial bases is sizeable. Iraq, Syria, and Jordan advocate, in varying degrees, self-reliance-oriented strategies, depending on their economic size, resource endowment and level of economic development. During the late 1970s, Syria and Jordan experienced high rates of growth in their Gross Domestic Products, which reflected both increased income and higher efficiency of investment. The Arab–Israeli war disrupted Jordan's and Syria's economies, but they have since achieved steady growth. Syria's economy was traditionally based on agriculture and trading; today almost 30% of its labor force is in the industrial sector. The Jordanians working in the rich Gulf region represent a major economic base to their country by transfers of their savings to help their home country offset its balance-of-payments deficit. Prior to the 1980 outbreak of war with Iran, Iraq's economy, based on oil, natural gas, and other minerals, showed a favorable trade balance.

Algeria, the North African nation in this classification, has some of the world's largest natural-gas reserves and is the continent's largest producer. Although about 40% of its people are engaged in agriculture, industry and primarily oil and natural-gas extraction dominate the economy.

LOWER-MIDDLE-INCOME COUNTRIES

Almost 60% of the entire Arab world and one-third of the Arab countries discussed here fall into a lower-middle-income economic bracket (annual GNP per capita: US $400–$1600). These relatively poor nations lag behind the other Arab countries. Except for Egypt, they lack a solid industrial base; and, despite partial industrialization, even Egypt's per capita income remains comparatively low. In the Sudan, agriculture dominates the economy, although only 5% of the land is farmed. The two Yemens have agricultural bases and lack mineral resources. For all three countries, the real obstacles to development of the agricultural sector include lack of transportation and communications infrastructures, land tenure and rapidly growing populations. Improving the infrastructures would require intensive capital, which they lack. All receive some economic aid from regional and foreign sources. The two Yemens could not exist without foreign support.

Morocco and Tunisia are more developed economically than the other three nations in this category. Their most important resources are minerals, notably phosphates, Morocco being the world's third-largest phosphate producer, after the USA and the USSR. Although agriculture remains important, their natural resources give Tunisia and Morocco an advantage which facilitates their economic development.

A social analysis

In recent decades, the Arab states have experienced accelerating social change. Arab scholars agree that the centuries-old traditional social structures of their nations are giving way to a new social order. Dynamic change is occurring, albeit at differing rates from country to country, on the social, economic and political levels. Kerr (Kerr & Yassin 1982) has called the massive migration of workers, technicians, and other experts to the oil fields the "engine" of change. Other scholars list additional forces leading to the new social order: increased emphasis on education, external and internal conflict, travel, trade, and impact of the outside world.

Social change began at the end of the 18th century when Western influences penetrated the Arab society. Ibrahim (1982) attributes much of the social changes to four forces: (a) colonial influences; (b) modern science and technology; (c) the struggle for independence; and (d) oil. With these impacts, the influence of traditional social orders diminished, but did not disappear. Instead, the traditional structures co-exist with modern concepts; their interplay has put the entire Arab world in a state of constant transition. Arab social change must therefore be seen as a process in ferment.

During the days of traditionalism, common traits characterized all the nation-states and sheikhdoms. Societies were patriarchal, patrilineal, usually patrilocal, and controlled by a small ruling class. Men dominated; women bore children and served as the backbone of close-knit families. Three levels of hierarchal family structure existed: the tribe, the clan, and the extended family. Stringent religious and moral codes prevailed.

Today, divergent political philosophies divide the Arab World. Some are oriented to the West, others to the East. Former "have-not" states have taken the ascendancy because of oil and mineral wealth. Education, much of it offered free by governments, has changed attitudes and produced modernizing trends. In a few countries, resurgence of religion has led to a return to traditional values; in others, loosening of religious ties has produced secularized societies. Movement to cities has impacted social structures and has brought modernization, but much of the region remains rural and strongly traditional.

Despite diversity, commonalities unite the region – a common language, a predominant religion, a rich literary heritage, a sense of continuing history as the "Arab nation," common patterns of family life, similar value and cultural orientations, economic complementarity, and, in most cases, sharing of independence, development, and recognition in the world family of nations.

Table 16.2 The Arab world: urban population growth.

Country	Percentage of urban population	
	1965 (%)	1983 (%)
Algeria	38	46
Egypt	41	45
Iraq	50	69
Jordan	47	72
Kuwait	75	92
Libya	29	61
Morocco	32	43
North Yemen	6	18
Oman	4	25
Saudi Arabia	39	71
South Yemen	30	37
Sudan	13	20
Syria	40	48
Tunisia	40	54
United Arab Emirates	56	79

Source: World Bank 1985.

The diversities and accelerating change make social classification difficult. However, three social groupings are emerging, all influenced by change. The "traditionalists" represent the element most resistant to change. The "technocrats" reflect the impact of outside forces, recently discovered resources and urbanization. The "entrepreneurs" manifest sensitivity to political ideologies and corporate business activity. These three groupings overlap considerably. Sometimes individuals or subgroups evidence characteristics of all three, but taken together, these groups comprise the "shapers" of a new Arab social order.

THE TRADITIONALISTS

For centuries, geophysical conditions divided Arab society into three major life-styles: (a) desert nomads; (b) farmers and villagers; and (c) urban dwellers. However, in recent decades, the discovery of oil, industrial and economic expansion, education, and foreign influences have speeded the urbanization and liberalizing tendencies which have changed the nomadic–rural–urban balance. Table 16.2 illustrates the accelerating urban growth. The most significant changes in life-styles and values took place in the cities. In the process, the gap between the traditional rural and modern urban life-styles widened.

Classical traditionalists, represented by nomads and farmers not highly exposed to modernization, resist social change. Nomadic Bedouin tribes still comprise over 10% of the population in nations with desert areas (e.g., the Gulf States). Similarly, farmers and villagers, especially the older generation, cling to strong emphases on the family, domination by the male, hard work, and traditional value systems.

Classical traditionalists also reside in urban areas. Often, they have come in search of work, bringing their conservative ways with them. They find reassurance by clinging to old values amid the confusion of liberalizing tendencies. But modernization affects most of them. Melikian & Diab (1983) found that traditional respect for the family was perceived to be the most important value among American University of Beirut students, with religious and ethnic mores second and third. In another study of social change in urbanizing Qatar, Melikian & Al-Easa (1983) concluded that current trends indicate that the nuclear family will replace the traditional extended family, that couples will increasingly reject arranged marriages, that endogamy will decrease, that families will have fewer children, and that women will strive for a more equal partnership in the family.

The neo-traditionalists place their faith in return to the historical achievements of Islam. Essentially, they are reactionaries who challenge secularizing influences. They revive Islamic symbols, sometimes militantly making them the framework of political opposition to the status quo. They embrace traditional life-styles. They have revived Friday attendance at the mosque. Among female college students in Egypt, for example, there has been a return to the veil and an Islamic dress code (Fernea 1985). El Guindi (1983) believes this neo-traditional surge represents a grass roots movement to revitalize Islamic values and to re-integrate Islam into a culture interwoven with people's daily lives, needs, and values.

THE TECHNOCRATS

The discovery and development of natural gas and oil reserves and industrial development in half the Arab nations have produced demands for a wide array of technical expertise. Workers from many nationalities have responded. Hundreds of thousands came from other Arab countries to work in the oil-producing nations. Others came from the West or Asia, especially India, Pakistan, and the Philippines. These immigrants brought differing life styles and liberalizing ideas with them. Sherbiny (1984) reports that the 1980 expatriate labor ratios in the top oil-producing countries were: Saudi Arabia, 53%; Kuwait, 79%; UAE, 89%; Iraq, 14%; and Libya, 34%.

The exodus to the oil states created labor shortages in some labor-exporting countries. Workers from other Arab nations filled the gaps. For example, in 1985 thousands of Lebanese, Palestinians, Egyptians and Syrians had joined 300,000 Jordanians working in the oil fields (Jordan Information Bureau 1985). Unskilled Arab replacements came principally from Egypt, the Sudan and the two Yemens. In general, all earn higher wages than at home. Many bring their families or send for them when they are established in their jobs. Most settle in the cities of their host countries.

This vast migration affects individual Arabs and the social and economic development of the entire region. Jordan's Crown Prince Hassan (Bin Talal 1984) cites five results: (a) the shifting demographic balances within Arab states; (b) the influx of influence from outside the Arab World; (c) the impacts of migrants on the host countries; (d) the effects of migration on the countries of the laborers' origin; and (e) the influence of uprootedness on the migrants themselves.

First, the heaviest influx of imported technocrats has occurred in the smallest states. The World Bank (1985) indicates over four of every ten persons in Kuwait, Qatar, and the UAE are non-nationals, causing strong implications for the economic growth and social transformation of these small countries.

Next, the ratio of non-Arab to Arab technocrats is increasing. In 1975, 65% of migrants were Arab; today, about half the technocrats are from outside the Arab world, two-thirds being from Asia, as LaPorte observes (1984).

Thirdly, the migrant technocrats are making a dramatic social impact on the oil-rich countries. The technocrats place numerous demands on host nation administration, social services and security, sometimes causing the restructuring of such services. Their presence sometimes raises questions about national security and political balance (e.g., Saudi Arabia and Kuwait). Since most migrants bring their families, the host nation's educational systems are often taxed beyond capacity and require expansion. The high proportion of expatriate children is certain to have a long-term effect on Kuwaiti schoolchildren. Ibrahim (1982) indicates that the migrants are virtually shaping the new institutions and infrastructures of some Arab countries. Bahry (1982) declares that the imported technocrats have expanded the horizons of Saudi families and have profoundly affected Arab life-styles. The modernizing influences introduced by technocrats erode traditional cultures by synthesizing traditional culture with the values introduced by the technocrats. Some changes are reflected in increased divorce, crime, and social drinking. Some persons appear to have adopted double moral standards.

As discussed in the "high income" section in the economic analysis above, technocrats also exercise strong influences on their countries of origin.

Finally, migration strongly affects the technocrats themselves. Attitudes toward work and money often change. Since nearly all receive more wages abroad than at home, consumption patterns change. Tastes acquired while away are often difficult to sustain upon return; dissatisfaction with the home country results. Many have learned to expect the state to provide education and social services, creating new dependences on their own government rather than on the individuals themselves.

In summary, technocrat migration erodes traditional culture, introduces new and diverse life-styles, liberalizes traditional moral and behavioral standards, and ushers in a broader "world view."

THE ENTREPRENEURS

Merchants comprise much of the economic strength of the Arab World. Since the development of oil and industrialization began, a socially significant, though numerically small, group of business "entrepreneurs" has arisen who wield power out of proportion to their numbers. They serve both as guardians of established values and as stimulators of interaction which facilitates growth and change. The entrepreneurs fall into three classes: (a) a growing and highly influential group of Arab government–multinational corporation go-betweens, which we label "capitalist entrepreneurs"; (b) a larger group of "sponsors" (Al-Kafil), who make middle-man arrangements, mostly in the private sector; and (c) the "enterprisers," who engage in indigenous creative business investment and expansion.

The first two classes are complementary, one working at high-government to conglomerate levels, and the other on the human resource plane, "enabling" government–corporation activity and the successful completion of development projects. Both groups are highly sensitive to governmental goals, political ideology, and corporate business activity. Because they hold key information or have influential personal connections, both serve as gatekeepers with a disproportionate influence on others. Both deal with foreigners. Both act as legal go-betweens. Both take personal risks, but usually invest little of their own money in their activities. Both understand clearly how their own society functions. Both keep abreast of political and social developments. Ibrahim (1982) calls both groups "lumpen capitalists" since they are neither fully productive nor parasitic citizens.

The capitalist entrepreneurs, usually educated in the West, begin by taking leadership in government or business, in the latter case often by taking up partnerships with foreign businesses or agencies. With a small capital outlay and usually with a few like-minded friends, they start their own corporation with many functions – consultancies, engineering, management, research, etc. Often they have relatives or friends in government. Usually they handle contracts worth millions. The capitalist entrepreneur performs valuable social services. He organizes, establishes contacts with the outside, and gets things done. He serves as a public relations diplomat on behalf of his nation, by interpreting his country to the outside and vice versa. As a shrewd opportunist, he assures that his dealings will bring handsome profits. Without his efforts, national development projects would move at a snail's pace.

The second class of entrepreneurs, the "sponsor" or enabler, usually comes from the working class. Often he has little education, but he is shrewd, sometimes tough. Typically, he arranges for the travel, employment, and oversight of expatriate workers or businesses. He holds legal responsibility both to government and to expatriates. He clears expatriate documents, travel, entry and residence permits, and uses the law to wield power over the migrants. Or he may sponsor and give legal cover for a new expatriate business, later claiming up to 50% as his cut. Finally, he travels abroad to select suitable expatriate labor or professionals. Sponsors serve two key functions: (a) they help to regulate the flow of desirable import labor and business, and (b) they protect the laws, traditions, life-styles, and cultures of their countries. Expatriates who do not heed their suggestions can be dismissed.

The third class of entrepreneurs, the "enterprisers," are alert Arab businessmen who introduce innovative ventures using indigenous resources. Their ideas may come from the outside, but they usually depend on indigenous ingenuity to make their ventures work. One such enterpriser has been Albert Abella, whose food catering services in Saudi Arabia have now expanded worldwide. The enterpriser may not be highly educated or trained in business administration, but he is creative, he is a wise investor, and he has seemingly innate sensitivity to the right time and place for expansion. The enterprisers make excellent contributions to advance.

Taken together, the three classes of entrepreneurs wield economic power disproportionate to their small numbers, they represent powerful social influences, and they contribute significantly to change and development.

Conclusion

Although economically diverse, all Arab governments have encouraged development and modernization. Oil, the "engine" of progress, has brought social changes, some of which threaten stability and security (e.g., in Saudi Arabia and Kuwait). The emphasis on development brings internal improvement; for example, the UAE uses oil revenues to encourage local agriculture (Richey 1982). On a one-to-one basis, the oil-rich nations have provided generous loans and grants to the less wealthy. Unfortunately, region-wide cooperation has not flourished.

During the recent growth years, expatriate workers have inflated the populations of the oil-producing nations. The production of oil has dramatically increased per capita GNP in the entire region, especially in oil-rich states. In the process, three new social groupings arose. Especially in the oil states, capitalist entrepreneurs facilitated needed aid and contractors' assistance to development, making a highly significant contribution to progress. Sponsors served a crucial social and economic rôle by regulating the flow of desirable labor to underpopulated countries with manpower shortages (e.g., UAE, Libya, Kuwait, and Saudi Arabia). Enterprisers, especially Lebanese,

Jordanians, and Palestinians, are increasing Arab self-sufficiency by developing indigenous resources.

However, the region faces four threats to continued rapid development – the oil glut, the Iran–Iraq and Arab–Israeli conflicts, and the need for better planning to transform assets into enduring developmental gains. Dropping oil revenues are decreasing demands for expatriate laborers and increasing threats to the stability of conservative regimes. Muir notes that plunging oil prices have helped rekindle Iran's military attacks on Iraq (Muir 1986). The continuing Arab–Israeli strife is costly, destabilizing, and divisive of Arab unity and progress. The economic slowdown and the conflicts are reducing the loan and grant aid that the oil countries can provide. Crucially-needed effective development planning is being postponed by current exigencies. These trends predict slowdown of economic development and social change for the region in the short run. However, the area's diversity and dynamism preclude dire predictions; throughout history, the Arabs have been resourceful and adaptive.

References and further reading

Abed, G. T. 1983. Arab oil exporters in the world economy. *American–Arab Affairs*, Vol. 3 (Winter, 1982–3), 26–40.

Ahrari, M. E. 1984. OPEC's widening window of vulnerability, *American–Arab Affairs*, Vol. 7 (Winter, 1983–4), 106–19.

Ayoob, M. 1981. *The politics of Islamic reassertion*. New York: St. Martin's Press.

Bahry, L. 1982. The new Saudi woman: modernizing in an Islamic framework. *Middle East Journal* **36**,4, 502–15.

Barakat, H. 1979. Arab Society: prospects for political transformation. In *The Arab future: critical issues*, M. Hudson (ed.), 65–80. Washington, DC: Georgetown University.

Bin Talal, H. 1984. Manpower migration in the Middle East: an overview. *Middle East Journal* **38**,4, 610–14.

Dempsy, M. W. 1983. *Atlas of the Arab world*. New York: Facts on File.

Dickman, F. 1984. Economic realities in the Gulf, *American–Arab Affairs*, Vol. 7 (Winter, 1983–4), 50–4.

El Guindi, F. 1983. The emerging Islamic order: the case of Egypt's contemporary Islamic movement. In *Political behavior in the Arab States*, T. E. Farah (ed.), 55–66. Boulder, Col.: Westview Press.

Farah, T. E. (ed.) 1983. *Political behavior in the Arab States*, Boulder, Col.: Westview Press.

Fernea, E. W. 1985. *Women and the family in the Middle East: new voices of change*. Austin: University of Texas Press.

Ghantus, I. T. 1982. *Arab industrial integration: a strategy for development*. London: Croom Helm.

Hudson, M. C. (ed.) 1979. *The Arab future: critical issues*. Washington, DC: Georgetown University.

Ibrahim, S. E. 1982. *The new Arab social order: a study of the social impact of oil wealth*. Boulder, Col.: Westview Press.

Jordan Information Bureau 1985. *The Hashemite Kingdom of Jordan: facts and figures*, p. 3. Washington, DC: Jordan Information Bureau.

Keely, C. B. & B. Saket 1984. Jordan migrant workers in the Arab region: a case study of consequences for labor-supplying countries. *Middle East Journal* **38**,4, 685–98.

Kerr, M. H. & El Sayeed Yassin (eds.) 1982. *Rich and poor states in the Middle East: Egypt and the new social order*. Boulder, Col.: Westview Press.

LaPorte, R., Jr. 1984. The ability of South and East Asia to meet the labor demands of the Middle East and North Africa. *Middle East Journal* **38**,4, 699–712.

Melikian, L. H. & J. S. Al-Easa 1983. Oil and change in the gulf. In *Political Behavior in the Arab States*, T. E. Farah (ed.), 41–54. Boulder, Col.: Westview Press.

Melikian, L. H. & L. N. Diab 1983. Stability and change in group affiliations of university students in the Arab Middle East. In *Political behavior in the Arab States*, T. E. Farah (ed.), 19–26. Boulder, Col.: Westview Press.

Muir, J. 1986. Iran's push into Iraq linked to plunging oil prices. *The Christian Science Monitor* **78**,70 (March 7, 1986), 9.

Nimatallah, Y. A. 1984. Economic Trends in the Gulf Countries and Their Implications for Relations with the West. *American–Arab Affairs*, Vol. 7 (Winter, 1983–4), 69–75.

Richey, W. 1982. Changing the face of the desert: farms irrigated with oil. *The Christian Science Monitor*, Vol. 75, No. 4, B8.

Serageldin, I., J. Socknat, J. S. Birks & C. Sinclair 1984. Some issues related to labor migration in the Middle East and North Africa, *Middle East Journal* **38**,4, 615–42.

Sherbiny, N. A. 1984. Expatriate labor flows to the Arab oil countries in the 1980s, *Middle East Journal* **38**,4, 643–67.

Tibi, B. 1983. The renewed role of Islam in the political and social development of the Middle East. *Middle East Journal* **37**,1, 3–13.

Todaro, M. P. 1985. *Economic development in the Third World*. New York: Longman.

World Bank 1985. *World development report 1985*. New York: Oxford University Press.

Zaghal, A. S. 1984. Social change in Jordan. *Middle Eastern Studies* **20**,4, 53–75.

Zurayk, C. K. 1979. Cultural change and transformation of Arab society. In *The Arab future: critical issues*, M. Hudson (ed.), 9–18. Washington, DC: Georgetown University.

17 *South Asia: a region of conflicts and contradictions*

B. L. SUKHWAL

The term "South Asia" also signifies the Indian Subcontinent, which covers 4,468,000 sq. km including India, Pakistan, Bangladesh, Sri Lanka, Nepal, Bhutan, and the Maldives. The subcontinent is bounded to the north by the Himalayas (a Sanskrit word meaning "abode of snow"), the loftiest mountains in the world, by the mountains of eastern Assam and Burma to the east, and by the Sulaiman and Kirthar ranges as well as the Thar Desert to the west; thus South Asia is a distinct physiographic unit. It is also a peninsula, jutting into the Indian Ocean, with the Bay of Bengal on the east and the Arabian Sea on the west. Therefore, South Asia forms an integral part of Mackinder's "World Island," that is, the Euro-African-Asiatic land mass, the most important single geographical unit in the world. Its relative location with reference to other political areas gives it immense geopolitical significance. Demographically, economically, politically, culturally, and strategically, South Asia is too important for the rest of the world to ignore. The western industrialized nations need its tea, jute, sugar, spices, minerals and precious metals, engineering goods, small arms, its culture and much more; conversely, South Asia needs technology, capital, manufactured goods, sophisticated military hardware, and to some extent food in return. The interdependence of the industrialized nations and South Asia is vital for the peace and prosperity of the region and the world at large.

The evolution of the South Asian region has a long history of over 4,500 years, an even longer history than the Indus civilization. Its early civilization, revealed in the cities of the Indus basin, Mohenjodaro, and Harappa, and dating from 2,500 BC, had contacts with the Mesopotamian civilization. The history of the region reveals an endless succession of invasions by various peoples, all contributing in some way to the cultural growth of the land, giving a complex character to its civilization. There are many languages, many religious groups, different forms of art and architecture, vast regional and economic differences, diversified political ideologies, and overall a sense of social stratification.

During the periods of British colonial rule, the subcontinent was politically fragmented and economically inviable to a degree that both distressed and challenged the new national leaders of various countries of the region. To Westerners it seemed that the nation was united, but in reality the general public was fearful of the British rulers, and as a result did not raise their voice except during the 1857 army revolt and the freedom struggle during the first half of the 20th century. The British ruled nine large provinces, and the rest of South Asia was controlled by 565 local Rajas and Maharajas who had to depend on Britishers for defense and advice. The British did, however, make some positive contributions to South Asian life, such as the development of systematic posts and telegraph services, networks of road and railroad transportation, irrigation canals, development of urban centers and port cities, industrialization on a limited scale, higher education for a chosen few, modern practices of medicine, steam navigation, the creation of a modern army, a Western-based legal system, and the system of administrative services.

At the eve of British withdrawal, the subcontinent was divided into India and Pakistan; the two wings of Pakistan (West Pakistan and East Pakistan), were separated by more than 1600 km of Indian territory. In 1971, after a bitter 14-day war, Bangladesh attained her independence from Pakistan. Sri Lanka gained her independence from the British in 1948. Surprisingly, postcolonial South Asia was divided into only three major nation states besides the two small mountainous countries of Nepal and Bhutan and the two island nations of Sri Lanka and the Maldives. In these South Asian states there are numerous regional, linguistic, religious, economic, and cultural differences. For example, in India there have been separatist movements in Punjan, Tamil Nadu, Assam and Northeast India, and Jammu and Kashmir, but the country has held together for 39 years. Pakistan also faced persistent separatist movements for Pakhtunistan and Baluchistan, but the nation remained intact since the independence of Bangladesh in 1971. In Sri Lanka, the majority of Singhalese Buddhists and the minority of Tamil Hindus are at odds, and the antagonism is continuing; however, the country is surviving these strains. Nepal has also experienced inside turmoils. In essence, all the South Asian nations have faced fissiparous tendencies but all states have remained intact with separate individual identities.

The natural environment

South Asia possesses physical as well as climatic extremes. The highest mountains in the world, the Himalayas, form the northern boundary while lowland Bangladesh and the river deltas face the constant threat of flooding due to the flat nature of the terrain. The variety of physical environment is evident from the snow-clad ranges of the Greater Himalayas to a belt of transitional foothills turning into a series of discontinuous longitudinal vales called "the dunes." South Asia is clearly set apart from the rest of Asia by a broad no-man's-land of mountains, jungles on the eastern frontier with Burma, ice on the Tibetan border, and desert as well as low mountains in the west. The Himalayas have created a linguistic and cultural divide, as well as a natural barrier between China and the USSR in the north and South Asia in the south. The region is accessible only through the Khyber and Bolan passes in the northwest. During spring and summer, these mountains provide valuable meltwater to the subcontinent when it is in great demand. Moreover, the cold Siberian winds are blocked by the mountains, and the climate of South Asia remains moderate and monsoon-dominated.

The Indus and Ganges plain area has a very high density of population, the nucleus of economic activities, the center of political power, the best agricultural land, rich alluvial soils, the densest network of rail and road transportation systems, and a concentration of irrigation canals and wells; the southern peninsula, on the other hand, contains almost all the mineralized areas of the region, including iron ore, coal, mica, manganese, bauxite, diamonds, gold, thorium, lead, zinc, uranium, etc. It is also the production center for cotton, peanuts, tropical fruits, and rice. The Deccan Plateau is the center of heavy industries including iron and steel, textiles, machine tools, shipbuilding, and transportation equipment (Fig. 17.1). The western and eastern coastal plains are important for production of petroleum (offshore and onshore) and various cash crops such as cashews, rubber, tropical fruits, coffee, spices, tea, coconut and coir, and rice. The Europeans first made contacts along these coastal plains at Cochin, Bombay, Madras, and Calcutta.

There are climatic extremes in South Asia: the temperatures in the Thar Desert reach over 50 °C in the summer months, in direct contrast to the shimmering, ice-capped Himalayas. The region receives nearly 90% of its rain during the southwest monsoon season from June to September. The rainfall is torrential in nature, resulting in excessive run-off and flooding, whereas too little rain creates conditions of drought and famine. The unreliability of monsoons in their onset and departure creates uncertainties for farmers. The unequal distribution of rainfall with an average of 11,430 mm at Cherrapunji to less than 25 mm in the Thar Desert creates fatalistic attitudes among the farmers. Water is a more valuable resource than land itself. Nowhere else are so many people so intimately dependent upon rainfall rhythms; the whole prosperity of the region is tied up with the eccentricities of its seasonal winds and rainfall.

Economy and development

At the time of independence in 1947, South Asia had a depressed economy and was considered an undeveloped region. The economic picture of the Indian subcontinent at the time of independence was grim. Barely 5% of the population earned more than enough for a subsistence living, and millions were unemployed. Problems such as rehabilitating displaced persons from

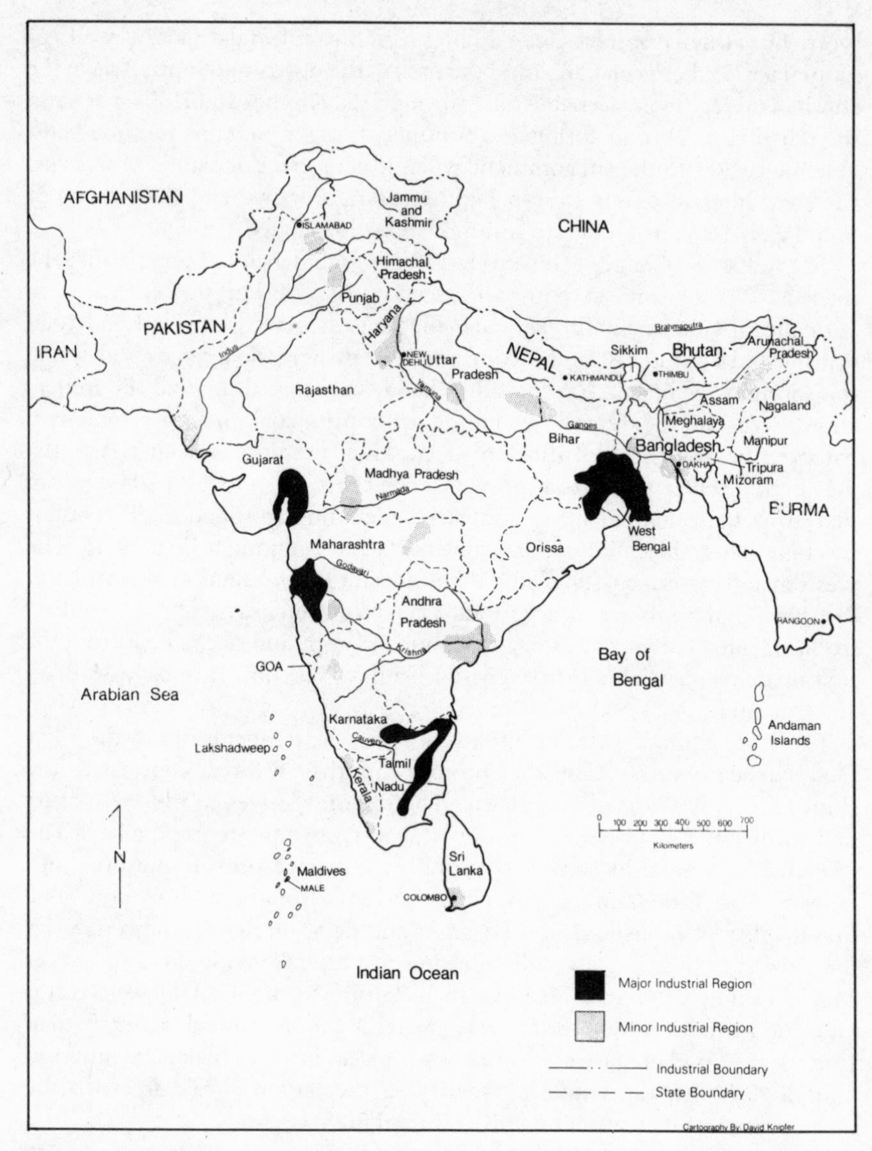

Figure 17.1 South Asia: industrial regions.

Pakistan to India and vice versa, establishing communal harmony, and increasing even slightly the economic level of the people, loomed as formidable for the newly independent countries of the region. Since independence, however, various countries have embarked on Five Year Plans (Sri Lanka had a Ten Year Plan). The utmost importance was given to agriculture, land reforms, mineral exploration, industrialization, and improving the transportation networks. Nearly 65% to 90% of the population has been directly engaged in agriculture, with nearly two-thirds of the land under food production, especially grains. South Asia feeds and clothes its population from less than an acre of land per person. There is probably no region in the world where land is so overworked. Until recently, food shortages were common because of disastrous weather conditions, including droughts, floods, famine, backward farming techniques, shortages of fertilizers, lack of irrigation facilities, limited and high-cost rural credit, and excessive pressure on the land by an ever-increasing population. The production per hectare is one of the lowest in the world. To improve the production per hectare, the "Green Revolution" was introduced in the 1960s and bore fruit in the 1970s. Through this revolution new strains of wheat (experimentally developed in Mexico) and rice (developed in the Philippines) were introduced into South Asia, requiring heavy use of fertilizers, irrigation, small machines, and new, as well as improved, seeds. Per hectare production increased sharply; the crop was called a "miracle crop." The result of the Green Revolution was that India produced a grain surplus with nearly 160 million tons of grains produced in 1985–6 with a surplus of 31.7 million tonnes. Pakistan is reaching the self-sufficiency level. Bangladesh, Sri Lanka, Nepal, and Bhutan may remain deficient states.

At the time of independence, less than 2% of South Asian workers found employment in industry, manufacturing and mining, producing less than 5% of the regional income. By 1985, industrial production had expanded seven-fold, especially in India. Highly sophisticated industrial products, at one time available only to consumers of industrial nations, are plentiful in South Asian markets. Through the planned industrialization, India has developed a high level of technological sophistication in such industries as textiles, sugar, iron and steel, machine tools, railway wagons and locomotives, automobiles, television sets, electronics, computer manufacturing, chemicals and explosives, cement, transportation equipment, aircraft manufacturing, shipbuilding, and domestic goods. New industries were established according to the availability of raw material and other industrial infrastructure. All regions were given equal consideration for industrial development and most of the states built industries appropriate to the area's resources. Currently, India is the third-largest producer of steel in the East and ranks seventh among the industrialized nations of the world. Its cotton textiles and tea, and several of its agro-industrial products, are among the best in the world; and the value added showed more than a 19-fold expansion from 1947 to 1985. The industrial development in the last 39 years has made India a power to reckon with in southern, southeastern, and southwestern Asia and in the world at large.

Industrial development in other South Asian countries has been slow and haphazard because the mineral resources are practically non-existent. Pakistan has some coal and natural gas and Bangladesh has some natural gas; thus

Table 17.1 Demographic, political, and economic data for South Asia.

	Area (1,000 sq. km)	Population 1985 (m)	Life expectancy (years)	Average annual growth rate (%)	Population density (No./ sq. km)	Form of government	Capital city	Per capita income ($)	GNP 1986	Annual growth rate: average GDP, 1980–4 (%)	Annual growth rate: agriculture average, 1980–4 (%)	Annual growth rate: industries average, 1980–4 (%)
India	3,204.1	770	54	2.2	214	federal republic	New Dehli	280	215,600	6.20	5.00	6.02
Pakistan	828.5	101	50	3.0	122	dictatorship	Islamabad	360	36,300	6.02	5.58	9.06
Bangladesh	144.0	104	47	3.1	722	dictatorship	Dapha	150	15,600	3.22	2.98	1.56
Sri Lanka	65.0	16.5	64	1.8	254	socialist republic	Colombo	320	5,280	5.30	3.88	3.92
Nepal	140.9	17.0	45	2.5	121	constitutional monarchy	Kathmandu	170	2,890	2.10	1.62	1.80
Bhutan	46.9	1.5	43	2.2	32	monarchy (Indian Protectorate)	Thimpu	90	135	8.05	–	–
Maldives	0.3	0.2	48	3.0	581	republic	Male	308	61.6	11.80	7.70	8.46

the major industries are agriculturally-based. Textile industries have improved to satisfy home demand in Pakistan. Other industries are chemicals, cement, bicycles, fertilizers, a small steel plant, an automobile plant in Karachi, and agricultural processing plants. Bangladeshi industries are cottage industries including textiles, farm implements, forestry products, fisheries, fertilizer plants, and food processing. The major industry in Bangladesh is jute processing. Most of Sri Lanka's industries are located in Colombo, the capital city; these industries include cement, shoes, textile, paper, china, glassware, and agro-industries. Nepal, Bhutan, and the Maldives have very little industrial development.

South Asia's biggest problem is the population explosion. With less than 2.95% of the total world area, it contains 20.4% of the population (Table 17.1). Every year about 20 million people are added to the region. Nearly one-third of the population faces the problem of malnutrition, and some of the people are unable to secure adequate food. There are still some deaths due to starvation, especially in Bangladesh. Some people live in abject poverty without proper sanitation, shelter and health facilities. The rapid growth rate also requires additional food, housing, educational facilities, clothing, and other requirements which the region cannot provide. To cope with the population problem, India, besides using other family planning methods, has started the massive family planning program under which over 34 million sterilizations have been performed between 1952 and 1985. It is estimated that between 1956 and 1983, 54.73 million births have been so averted in India, a phenomenal achievement in a developing country. Still the program needs constant encouragement, and the birth rate has to be lowered to achieve a zero growth rate in population. Pakistan and Bangladesh, however, have not attempted family planning programs on a national scale like India.

Development strategies

Since independence, most of the South Asian countries embarked on planning processes of one sort or another. An extensive planning program was introduced by India, and the Government of India appointed a Planning Commission in 1950 to prepare a blueprint of development, taking an overall view of the needs and resources of the country. The First Five Year Plan (1951–2 to 1955–6) accorded the highest priority to agriculture, including irrigation and power-production projects. The Second (1956–7 to 1960–1) and Third (1961–2 to 1965–6) followed the same pattern, with maximum emphasis on agriculture and basic industries. The planning process, industrialization, and agricultural development have not produced balanced regional development, each state competing for resources from the center. The first three five-year plans were biased towards the urban areas, so the Fourth Five Year Plan (1968–9 to 1973–4) laid more emphasis on improving the conditions of less privileged and weaker sections of the society. The program, however, was carried out by different agencies and, as a result, lacked proper coordination and implementation. The macro- and meso-level planning was unrealistic and increased income disparities and unemployment. To correct the imbalances of development and lay more emphasis on rural development, a program of Integrated Rural Development was prepared after the Fifth Five Year Plan (1974–5 to 1977–8) to integrate and coordinate various development programs in rural areas, i.e. to form a single strategy for development. It was based on the principle of all-round development of the rural areas through micro-level planning which included agriculture, industry, transport, commerce, health, education, market centers, urban development, and the balanced development of all regions. This process involved multi-level planning including the village, village Panchayat Samitis, blocks, tehsils, districts, state, and the country. The same program continued during the Sixth Five Year Plan (1980–1 to 1984–5); however, the goals of removing poverty and building a modern society were not achieved. Nearly 273 million people (37%) remained poor.

An ambitious Seventh Five Year Plan (1984–5 to 1989–90) was launched by the new Prime Minister, Mr. Rajiv Gandhi, as part of the long-term strategy virtually to eliminate poverty and illiteracy, achieve near full employment, secure satisfaction of the basic food needs, clothing and shelter, and provide health care for all by the year AD 2000. In absolute terms, the number of poor persons is expected to fall from 273 million in 1984–5 to 211 million in 1989–90. The new planning strategy has invited large-scale private investment, more so than in previous five-year plans. How far India will be able to succeed in achieving her goals is uncertain, but at least she has regularly carried on a strong viable planning program.

Development programs in Pakistan were initiated in 1955 with its First Five Year Plan, with a much greater rôle to the private sector in agriculture and industries than in India. The Pakistan Industrial Development Corporation established industries that were handed over to private firms as they became available. The development program was export-oriented and relied heavily on foreign expertise and financial aid. During Bhutto's rule, Islamic socialism and the nationalization of industries were initiated and radical land reforms were to be implemented, but he was overthrown in 1977; afterwards the earlier strategy was restored. Pakistan's development program depended more on Middle East and American aid and has not increased productivity and employment or the equitable distribution of resources and income.

Since the independence from Pakistan in 1971, Bangladesh set up a Planning Commission in 1973 to create a mixed economy and allow for a transition to socialism. The plan laid heavy emphasis on agriculture and on employment. Industries were nationalized. After the coups in 1975, socialism was no longer seen as the ultimate goal. Instead, new measures on denationalization were introduced and agrarian reforms were proposed, including a "due share of crops to the landless." This plan did not succeed either. The long-delayed Second Five Year Plan (1980–5), with the objectives of expansion of employment, assurance of basic needs, equitable income distribution, comprehensive rural development, and self-sufficiency in foodstuffs, was introduced. This has also not raised the standard of living.

Sri Lanka's First Ten Year Plan, prepared in 1959 by the Sri Lanka Freedom Party, proposed to achieve suitable industrialization progressively by importing capital goods and exporting the diversified agricultural produce as well as precious metals. In 1957 all foreign-owned estates were nationalized so that Tamil landlords would not be able to transfer their resources to India. When the United National Party came to power in 1977, a Free Trade Zone was established in an effort to attract foreign capital on the capitalistic model of Singapore. This policy is still followed by the Planning Commission. The Planning Commission is more active on land reforms and irrigation projects which have helped the island nation in increasing its cash-crop production and the growing of rice as a main food source. Nepal has followed the model of India in its planning programs; however, due to lack of investment and a poor natural resource endowment, the country has not made much headway on the planning and development front.

The planning process in South Asia constitutes a major part of economic development, but it has not been a balanced one. Except for India, the Five Year Plans or Ten Year Plans have not been administered satisfactorily and have not produced the desired results. Even in India, where the planning process has been coordinated between the center and the states, achievements lagged behind targets and development has produced regional imbalances and has increased the gap between the rich and the poor. Overall, the governments of South Asia may continue such planning processes. Future development is likely to be more balanced than it has been during the past 35 years.

Cultural and political patterns

The history of the Indian subcontinent reveals an endless succession of invasions by various peoples through the Khyber and Bolan passes and later by sea, resulting in complex cultural patterns of linguistic, religious, cultural, economic, and political diversities. The internal boundaries of India have been delimited on the basis of languages, and the nation has 14 major languages and 865 dialects and tongues. Similarly, Pakistan has the Urdu, Punjabi, Sindhi, Pakhtoon, Baluchi, and Pahari languages; whereas the people of Bangladesh speak Bengali and Urdu. Sri Lanka adheres to Singhalese and Tamil as the major languages. People of Nepal and Bhutan speak Nepali and Dzongkha, respectively. A minority of people in these countries who are university-educated also speak English as a second language.

The subcontinent is also divided on the basis of religions. Pakistan adopted fundamental Muslim law and Bangladesh is also Muslim. Sri Lanka adheres to Buddhism and has Hindu and Christian minorities. Nepal is the only Hindu country in the world whose state religion is Hinduism. India, however, is a secular nation, incorporating all religious groups including Hindus, Muslims, Sikhs, Jains, Parsees, Buddhists, and tribal groups. Distinct cultural patterns in South Asia are also evident where the caste system is still practiced in everyday life, especially in rural villages throughout South Asia.

Politically speaking, Nepal and Bhutan are monarchies, whereas Pakistan and Bangladesh are military dictatorships. The Martial Law Administrator of Pakistan, General Zia Ul Haq, has recently declared that there will be civilian elections soon. Sri Lanka has a presidential form of government in which the president is elected for seven years. The Maldives has an elected parliamentary form of government. India constitutes the world's largest and most complex federal state, with a parliamentary form of government. Since independence, eight general elections have been held that were fair and free. It is the largest democracy in the world, with more than 400 million voters. While many countries in the Third World and even some in Europe and Latin America have adopted other forms of government, India, in spite of the enormity of its social, economic, linguistic, religious, and regional problems, has remained wedded to the parliamentary form of government and democratic values. India has proved that successful changes can be brought about in a developing country through peaceful and democratic means.

International relations

Geopolitically, South Asia will doubtless remain among the world's most politically volatile areas and of great strategic importance to world powers. The region borders the Soviet Union and China in the north; these ideological foes and the United States hold naval bases in the Indian Ocean. South Asian nations follow neutrality, except Pakistan, which receives military aid from the United States. India is an active member of the non-aligned movement and in fact has been the chairman of the group. The 120 member nations occupy over 36% of the land area and 42% of population, with a majority of the nations from the Third World. Pakistan has been allowed as a sitting member, while the rest of the South Asian nations are regular members of the group. Peace and stability in Asia, particularly in South Asia, are matters of great importance to all nations of

South Asia and the world at large: if the Indian subcontinent remains free of tension, it could command unique weight among the community of nations. The Prime Minister of India, Mr. Rajiv Gandhi, has taken an active interest along with five other world leaders not only in reducing nuclear weapons but in eliminating them from the face of the earth. Since coming to power with a massive majority in the parliament in the eighth general elections in December 1984, he has initiated the South Asian Association for Regional Cooperation (SAARC), an organization of all South Asian countries pledged to cooperation in matters of agriculture, meteorology, tourism, and sports; to non-interference in each other's international affairs; and to peaceful settlement of all outstanding regional disputes. India and Pakistan fought three wars over Kashmir and the matter still lingers on; however, the rival nations are coming closer and have decided to settle all disputes by mutual agreement through peaceful means. This is an obvious sign of the establishment of peace in the region.

In summary, the conflicts and contradictions of South Asia are evident in all physical and cultural spheres. Physiographically, the north is sheltered by the high Himalayas and the south by the old Deccan Plateau, both separated by the flat Indo-Gangatic plain. South Asia experiences 50 °C in the Thar Desert to below-freezing weather in the Himalayas. Similarly, Assam and western Ghats receive very heavy precipitation while the Thar gets practically no rain. The east coast of India and Bangladesh experiences cyclones and flooding, and during the same period some parts of central India face the specter of drought. Nepal and Bhutan regularly face the horrors of landslides, whereas Sri Lanka and the Maldives are cut off from the mainland as islands. The environment is at times a friend and a foe alike.

In the north, one finds that the oldest civilization centers of Mohenjodaro and Harappa, the Aryans; in the south, the Dravidians. Pakistan and Bangladesh adhere to a Muslim fundamentalist religion whereas Nepal is a Hindu state. Sri Lanka is predominately a Buddhist nation; on the other hand, India follows secularism with the majority consisting of Hindus. There is a range from strict fundamentalists to atheists in South Asia. Culturally, there are caste differences among Hindus, but some Muslims and Christians also follow caste tradition. There are large urban centers like Calcutta, Bombay, Dakha, Karachi, and Delhi; and, in contrast, countless small hamlets and rural nucleated villages. The population densities at places vary between more than 650 persons per sq. km in the rural areas of the Ganges delta and Kerala and less than 3 persons per sq. km in the Thar desert and the Himalayas.

South Asia also possesses some of the best agricultural land and irrigation facilities, in the Indus–Ganges basin, and also barren desert in the northwest. It contains the very highly industrialized regions of Chota Nagpur Plateau, Bombay, and Madras, and the completely rural agricultural regions of Bangladesh. The region has followed varied development strategies, with

India committed to improve its economy through planning processes, whereas Bangladesh and Pakistan have used planning haphazardly and without proper coordination. The region has the landlocked countries of Nepal and Bhutan and the island nations of Sri Lanka and the Maldives. It contains the small nations of the Maldives and Sri Lanka and the large nation states of India and Pakistan. Politically, South Asia ranges from the strong military dictatorships of Pakistan and Bangladesh to the largest democracy in the world, India, to the monarchies of Nepal and Bhutan.

South Asia is thus a museum of races and cultures, cults and customs, faiths and tongues, with a vast gap between the rich and poor and between the educated and the illiterates. Yet despite this diversity there is a considerable degree of unity as to common historical bonds, cultural similarities, economic necessities, and common interests of the region bind all these nation states together.

Further reading

Bhardwaj, S. (ed.) 1983. *Hindu places of pilgrimage in India: a study in cultural geography*. Berkeley: University of California Press.

Chakravati, A. 1973. Green revolution in India. *Annals of the Association of American Geographers* **63**, 319–30.

Dutt, A. *et al.* 1972. *India: resources, potentialities and planning*. Dubuque, Iowa: Kendall/Hunt.

Farmer, B. H. 1983. *An introduction to South Asia*. New York: Methuen.

Hall, A. 1981. *Emergence of modern India*. New York: Columbia University Press.

Johnson, B. L. C. 1982. *Bangladesh*, 2nd rev. edn. Totowa, NJ: Barnes & Nobles.

Johnson, B. L. C. 1979. *India: resource and development*. Totowa, NJ: Barnes & Nobles.

Karan, P. P. 1960. *Nepal: a cultural and physical geography*. Lexington: University of Kentucky Press.

Karan, P. P. 1967. *Bhutan*. Lexington: University of Kentucky Press.

Miller, E. W. 1962. *A geography of manufacturing*. Englewood Cliffs, NJ: Prentice-Hall.

Myrdal, G. 1968. *Asian drama: an inquiry into the poverty of nations*, 3 vols. New York: Pantheon.

Noble, A. & A. Dutt (eds.) 1982. *India: cultural patterns and processes*. Boulder, Col.: Westview Press.

Nyrop, R. F. 1984. *Pakistan: a country study*. Washington, DC: US Government Printing Office.

Schwartzberg, J. (ed.) 1978. *An historical atlas of South Asia*. Chicago: University of Chicago Press.

Sopher, D. (ed.) 1980. *An exploration of India: geographical perspectives on society and culture*. Ithaca, NY: Cornell University Press.

Spate, O. H. K. & A. Learmonth 1971. *India and Pakistan: a general and regional geography*. London: Methuen.

Sukhwal, B. L. 1971. *India: a political geography*. New Delhi: Allied Publishers.

Sukhwal, B. L. 1976. *South Asia: a systematic geographic bibliography*. Metuchen, N.J.: Scarecrow Press.

Sukhwal, B. L. 1985. *Modern political geography of India*. New Delhi: Sterling Publishers.
Sukhwal, B. L. 1986. *India: economic resource base and contemporary political patterns*. New Delhi: Sterling Publishers & New York: APT Books.
Taylor, A. (ed.) 1974. *Focus on South Asia*. New York: Praeger/American Geographical Society.

18 *East Asia: contrasts in progress in a divided cultural realm*

THOMAS D. ANDERSON

No account of the Third World is complete without consideration of East Asia. Home for a quarter of the human race, it is the most populous of all world regions, yet only eight political units are present. Most land and people are part of the People's Republic of China (PRC; China). The other entities are the People's Republic of Mongolia, the Democratic People's Republic of Korea (North Korea), the Republic of Korea (South Korea), the Republic of China (Taiwan), Hong Kong, and Macao. The last two are colonies of Britain and Portugal respectively. Mongolia and Macao receive scant consideration here, Mongolia because its close political and economic integration with the Soviet Union makes it a virtual appendage of that state, and Macao because of its tiny size.

East Asia lies in middle latitudes on the southeast part of the world's largest land mass. This continentality adjacent to the warm Pacific Ocean produces seasons with contrasts and often extreme conditions. In winter air masses from the interior bring cold and dryness. Conversely summers are hot and oceanic air brings moisture. Precipitation decreases to the north and west, and the interior is desert. Winter precipitation is common only in coastal south China, Taiwan, and Japan.

Most of East Asia is hilly or mountainous. Other than the North China Plain and central Manchuria, the lowlands are small and fragmented. The combination of dryness and highlands reduce greatly the proportion of land favorable for agriculture, and for commerce and industry as well. The close association of high human densities with humid lowlands has produced an uneven population distribution.

*The Chinese cultural realm**

The main culture traits evolved very early in the middle Hwang Ho valley and diffused throughout the defined region. Practices definitive of the

* "Cultural Realm" refers to a large area where a unity of various traits distinguishes it culturally from other parts of the world.

cultural realm include the use of chopsticks, widespread use of human waste as fertilizer, and an ideographic alphabet. Racially all inhabitants except for several isolated peoples are classed as mongoloids.

Several non-material cultural attributes have affected modern development. Religious beliefs are mixed, with Buddhism the most widespread. Ethics more than laws provide societal guidelines. Temporal power has long been pre-eminent over religion. Traditionally the extended family was the base element of society, the center of personal loyalty, and the production unit in agriculture. Women held inferior status. Care of the young, sick, and aged was the responsibility of the family, with little recourse to outside agencies. Education held high status even where literacy rates were low.

Despite much provincialism, strong central governments fostered widespread national identification. Just who was Chinese, Korean, and Japanese was established early. The antiquity of the native civilizations has given East Asians keen appreciation of their past and a sense of self-worth in relations with outside peoples. Manifestations of colonial mentality are uncommon. The confidence has eased the adoption of foreign technology, which has been accepted and modified to local conditions with little xenophobic resistance.

MODERN HISTORY

Outside influences have contributed to contemporary political divisions. They began with the establishment of the Portuguese colony of Macao in 1557. The British acquired Hong Kong in 1841 and leased peninsular Kowloon for 99 years in 1898. In 1949 the triumph of communist forces under Mao Ze-dong over the Nationalist government brought about the PRC on the mainland and the Republic of China on Taiwan. Military backing by the United States helped maintain the circumstance of a divided China. At the end of World War II Korea was occupied by US and Soviet armies, which divided it south and north, set up friendly governments in Seoul and Pyongyang respectively, and then withdrew. The North's attempt in 1950 to reunite the nation militarily caused a bloody conflict that involved troops from China on the side of the North and those of the United States and many Western states in support of the South. A truce in 1953 left the two territories little changed in area and a residue of hatred that fuels intense competition between the compatriot states.

After 1945 United States occupation forces supervised a political restructuring of Japan which produced a democratic state within the context of an indigenous cultural and economic system. The result is a demilitarized Japan which is the world's second-ranking industrial power, a major innovator of technology, and the only Asian member of the First World. Japan serves as a growth model and investment source for a number of non-communist Asian countries. Its rôle is an inescapable element in the assessment of economic development in the region.

The Third World in East Asia

Neither China nor North Korea is part of the Second World, despite borders with the Soviet Union and communist governments. The PRC's close initial relationship with the USSR ended in the mid-1950s over issues both nationalistic and ideological. Old boundary disputes provoked armed clashes and continued confrontation, until 1987 when Gorbachev began a rapprochement. Although on better terms with the Soviets, North Korea has pursued an independent path in national and foreign affairs. These two countries, along with South Korea, Taiwan, and Hong Kong, are members of the Third World.

In essence the Third World in East Asia consists of two nations – Chinese and Korean – divided politically after World War II. Present are inner, or continental, countries and outer, or marine, countries; they have contrasting governments and economic systems. The inner group is favored by mineral resources, and by energy sources in particular though only China has petroleum. None of the outer countries has a strong mineral base and all rely on imports.

Thus the divergent development strategies reflect not only ideology but geography as well. In China and North Korea industrial programs were designed to use domestic resources to serve domestic markets. In accordance with conventional marxist practice, national "needs" were defined by central planning councils, not by popular demand. All aspects of society were embraced by centrally-supervised, multi-year plans that featured assigned achievement quotas. North Korea in particular strove for near autarchy under an official dogma termed *juche* (McCormack & Seldan 1978, p. 123). Both governments mobilized and relocated its peoples at will.

Periodically the PRC has instituted programs disruptive of its society and economy. In the name of the revolution collectivization in the early 1950s served to destroy traditional structures and entrench Communist Party power. The Great Leap Forward of 1958 aimed at instant progress and its abject failure hampered development for years afterward. The primary motive of the Cultural Revolution was to inculcate revolutionary fervor as the primary motivation for human endeavor. It was begun in 1966 and whimpered to an unsuccessful close with the death of Mao a decade later. During its turmoil several famines occurred.

New leadership under Deng Xiaoping installed in 1979 a national program based on material incentives. Its most important feature in a population 79% rural was a shift in production responsibility from collectives back to family units (Pannell 1985, pp. 170–1). Rigid central controls also were eased in commercial and industrial activities. Noted for the homily "Black cat or white cat, a cat is a good cat if it catches mice," Deng has directed a dramatic resurgence of the Chinese economy. The most evident change was in agriculture where grain production rose from 276.4 million tonnes in 1978 to

369.5 million tonnes in 1984. Although imports of wheat continued, by 1985 China competed with the United States in the export of maize to Japan.

The introduction of market incentives and greater management freedom within the context of a Leninist political system has been least successful in industrial sectors. The easing of import barriers brought a flood of better-quality foreign goods to a suddenly more prosperous people with a pent-up desire to acquire. Efforts to balance this capital flow through the export of Chinese products in competitive world markets and the creation of special enclaves for foreign investment were only partly successful. Such problems notwithstanding, Chinese economic growth in 1985 exceeded 11% and was the highest in East Asia.

These post-1979 policies have affected Chinese society as greatly as the campaigns of Chairman Mao. Different, however, has been the absence of proscribed ideological content and a need to orchestrate popular support. Public enthusiasm for the changes appeared to be genuine. Even so, the eventual success of this latest strategy was unclear in the late 1980s. Despite increases in the per capita GNP, China remained a poor country. Also, the reappearance of sharp income disparities among two generations of people accustomed to the barracks-like egalitarianism of Mao caused social strains that may yet induce political reactions by the Leninist power structure.

Hong Kong has been the main interface of centrally-planned and market economies. As such it attracted several million Chinese refugees seeking greater freedom and economic opportunity. With little space and fewer resources its assets have been accumulated capital, access to China, a good harbor, entrepreneurial skills, docile, low-cost labor, and an environment of freedom. They have been sufficient to generate a high rate of economic growth, close trade ties along the Pacific Rim, and a per capita GNP twenty times that of China and twice that of Taiwan.

Its economic future, however, is cloudy. When the British lease expires in 1997 all Hong Kong will come under Chinese rule according to agreements signed in 1984. Statements by Beijing that it will honor a policy of "one government, two systems" have reassured few Hong Kong residents. This skepticism has caused a decline in investments and a major capital outflow. As the reversion date nears a large increase in emigration also is probable. Political control by Beijing is expected to eliminate a zone of sharp economic contrast and cause a costly blow to the Chinese economy which has benefited greatly from the presence of Hong Kong.

Under the Nationalist government, the economy of Taiwan has consistently been export-oriented. Aid from the United States added to capital brought in 1949 from the Mainland funded industrial expansion. Foreign investments and labor-intensive branch plants of multi-national companies were solicited aggressively. The authoritative government supervised a docile, hard-working, low paid workforce and a stable political environment. Production of lower-priced consumer items and textiles has been emphasized, although since 1970s considerable expansion has occurred in the chemical, shipbuilding, and electronics industries. By 1990 the exports will include automobiles. Rural development has been widespread as well. There the quality of life, application of technology, and crop yields rival those of the First World.

Because of their many similarities the two Koreas offer an excellent basis for comparison of differing development strategies. Both states suffered severe damage to their economic infrastructure during the 1950–3 conflict. Under its zealous communist government North Korea rebuilt more quickly and by 1965 was clearly superior economically to the more populous South. However, from 1965 to 1975 while the North doubled its real GNP, that of South Korea tripled and surpassed the GNP of the North for the first time in Korean history (CIA 1978, p. 1). Until the world recession of the early 1980s South Korea had the fastest-growing non-OPEC economy in the Third World (Krause & Sekiguchi, 1980, p. 17).

As in Hong Kong and Taiwan, growth in South Korea was spurred by foreign investments, labor-intensive, consumer-oriented manufactures, and world markets. These characteristics differed in all aspects from those of North Korea. South Korea also earned foreign currency by the recruitment of construction workers for overseas projects, especially in Persian Gulf countries. By the 1980s a more balanced industrial structure made South Korea a world competitor in the sale of chemicals, steel, ships, and – in 1986 – of automobiles.

In both Korea government investments and self-help programs have raised farm income near to the levels in urban areas. Living conditions there have been improved with better roads, housing, electrification, and schools. Levels of mechanization and fertilization are high, as are crop yields and farm labor efficiency. Grain output was nearly equal in each country, amounts that made North Korea self-sufficient but which provided only 75% of the needs of the more populous South where imports made up the difference. However, maize was a larger proportion of the harvest in the North and because rice is the preferred food everywhere, the perceived quality of diet was higher in the South.

Militarization

East Asia is highly militarized. Even China's demobilization of a quarter of its 4,000,000-man People's Army in the mid-1980s did little to aid the economy as the savings were expended in modernizing its armaments. Proportionate to their populations, Taiwan and both Koreas have large military establishments. The justification for constant readiness in Taiwan is the official perception of an invasion threat from the Mainland. Since the 1953 truce both Koreas have remained highly mobilized and have actively confronted one another with over a half million men each. Estimated military expenditures were 15–20% of the GNP of North Korea and 5% of the larger GNP of the South. In addition, about 50,000 American troops were stationed in the South. A garrison-state mentality affects many aspects of life in Taiwan and the Koreas.

Demographic conditions

Because of its huge population, attention worldwide has focused on demographic circumstances in East Asia where data indicate that a demographic transition is in progress (Anderson 1980). It began first in Japan where the birth rate was halved between 1947 and 1957. Population growth continues in Japan but at a pace more subject to accommodation by social and economic planners.

In Taiwan and South Korea concern about rates of increase was expressed in the early 1960s. A program of action in Taiwan achieved a planned reduction of population growth rate from 2.9% to 1.9% by 1975. In South Korea the rate declined from 2.9% to 2.0% between 1960 and 1970. The pace of decrease then slowed greatly in both countries but the growth rates were viewed by authorities as manageable. The countries stand as examples wherein rapid modernization widened educational and economic opportunities which in turn influenced family decisions to limit family size.

Official policy shielded demographic conditions in North Korea from outside scrutiny. The data used on Table 18.1 are impossible to confirm. Indeed, one authority (Nick Eberstadt at the Harvard Center for Population Studies) estimated that infant mortality rates there ranged in the mid-40s rather than the low 30s. The demographic circumstances appear to be comparable to but less advanced than those in South Korea, although official incentives for higher birth rates were in place (Scalapino & Kim 1983, p. 120). Regardless, a steady decline was believed to be taking place.

Its great size makes the demographic situation in China the most significant although here also reliable data are scant. The lack is due not to official secrecy but to enormous numbers in a land where vital statistics were not recorded uniformly. National censuses were taken in 1953, 1964, and 1982, but only the last employed scientific techniques (Ma 1983, pp. 198–202). Thus comparisons of apparent trends require caution.

Promotion of a one-child policy in the 1970s caused the rate of increase to drop to about 1.0%. Official measures toward that end included pressures to delay the marriage age, the distribution of contraceptives, and increased numbers of sterilizations and abortions. Early in the 1980s Steven W. Mosher, an American anthropologist who lived for over a year in a Chinese village, published reports of forced abortions and female infanticides. The images stirred controversy worldwide and worsened relations between the

Table 18.1 East Asia demographic data.

Country	Area (sq. km)	Total population (million)	Crude birth rate (per 1,000 population)	Crude death rate (per 1,000 population)	Annual increase (%)	Projected doubling time (years)	Infant mortality rate (per 1,000 births per annum)	Population under 15/over 65 years (%)	Life expectancy (years)	Urban (%)	Per capita GNP (US $)
China	9,559,690	1,042.0	19	8	1.1	65	38	34/5	65	21	290
Hong Kong	1,062	5.5	15	5	1.0	67	9.9	24/7	76	92	6,000
Japan	377,389	120.8	13	6	0.6	110	6.2	22/10	77	76	10,100
Macao	2	0.3	25	7	1.8	38	38	31/5	68	97	2,560
Mongolia	1,564,360	1.9	34	7	2.7	26	50	41/3	65	51	–
Korea, North	120,539	20.1	31	7	2.3	30	32	38/4	65	64	950
Korea, South	98,485	42.7	23	6	1.7	41	29	32/4	66	57	2,010
Taiwan	35,988	19.2	21	5	1.6	44	8.9	31/5	75	71	3,000
Total	11,757,515	1,252.5									

Source: Adapted from Population Reference Bureau 1985.

United States and China. A reversal of the birth-rate decline after 1980 has been reported, presumably because designation of the family as the basic agricultural unit again made more children an asset. Regardless, the rate of population increase in East Asia was the lowest in the Third World (Aird 1985; Hsu 1985, p. 242).

Politics, culture, and progress

A striking feature of the related processes of demographic transition, social change, and economic development in East Asia is that they have taken place nearly simultaneously under a variety of political conditions. Japan was a democracy with much individual freedom. As colonies, Hong Kong and Macao had limited self-government but provided civil liberties comparable to those in the metropolitan states. In Taiwan and South Korea long-repressive right-wing governments moderated in 1987 and allowed active political opposition. Indeed, the ruling party in South Korea won only a plurality of votes in the December 1987 national election. Marxist–Leninist regimes ruled in Mongolia, China, and North Korea. These were totalitarian systems in which centralized authority not only prohibited all political opposition but also intruded into many aspects of personal life. The 1979 reforms in China, however, lessened controls on private enterprise, religion, and foreign travel. On the other hand the North Korean regime remains as repressive as any in the world. Participation in state-run organizations is mandatory and tendencies toward individualism are methodically smothered (McCormack & Selden 1978, p. 132). President Kim Il-sung is glorified as the embodiment of the Korean nation, with his son groomed as his successor.

This ideological diversity notwithstanding, major progress towards modernization is occurring throughout East Asia. The search for an explanation calls attention to cultural affinities of regional compass. Common societal qualities that fostered development are adaptability, discipline, industriousness, and a zest for education, all within an atmosphere of political stability. If valid, this interpretation raises doubts about efforts to posit as Third World development models either the central-planning policies of inner East Asia or the export-oriented systems of the outer countries. Neither approach is likely to work as well for peoples with different temperaments. In the absence of a clear answer to this conundrum, the evidence from East Asia seems to weaken arguments in favor of any single political route into a modernized future.

References and further reading

Aird, J. 1985. China must adapt to only child. *Wall Street Journal*. July 15, 13.

Anderson, T. D. 1980. Revolution without ideology, demographic transition in East Asia. *Philippine Geographical Journal* **20**, 33–44.

CIA (Central Intelligence Agency) 1978. *Korea: the economic race between North and South.* A research paper. ER 78–10008. January. Washington, DC: DOCEX, The Library of Congress.

Hsu, M. 1985. Growth and control of population in China: the urban–rural contrast. *Annals of the Association of American Geographers* **75**, 241–57.

Krause, L. B. & S. Sekiguchi (eds.) 1980. *Economic interaction in the Pacific Basin.* Washington, DC: The Brookings Institute.

Ma, L. J. C. 1983. Preliminary results of the 1982 census of China. *Geographical Review* **73**, 198–210.

McCormack, G. & M. Selden (eds.) 1978. *Korea North and South.* New York: Monthly Review Press.

Mosher, S. W. 1983. *The broken earth: the rural Chinese.* New York: The Free Press.

Pannell, C. W. 1985. Recent Chinese agriculture. *Geographical Review* **75**, 170–85.

Population Reference Bureau 1985. *1985 World population data sheet.* Washington, DC.

Scalapino, R. A. & J. Kim (eds.) 1983. *North Korea today: strategic and domestic issues.* Research Monograph. Berkeley, Cal.: Center for Korean Studies.

19 *Development in subSaharan Africa: barriers and prospects*

HAL FISHER

Introduction

SubSaharan African nations face a formidable array of impediments to their development. This chapter briefly describes the clusters of problems which slow their development and considers policies and potential for surmounting the difficulties.

The subSaharan area we consider extends from below the northern tier of Arab nations southward to South Africa and includes only the independent states of the region. The area falls between the Tropics of Cancer and Capricorn and extends north–south nearly 6,000 km and east–west some 7,000 km at its widest point.

Regional features

Topographical and climatic conditions play a crucial rôle in the region's fragile ecology. Large expanses are "bush" country or mixed grasslands and woods, highly dependent on a delicate balance of wet and dry seasons. The desert wastelands of the north and southwest are expanding, while the humid tropical forestlands are shrinking. Plateaux dominate much of the terrain, with rugged mountain systems in the east-central sector. Soils range from arid desert to marginally fertile; some subsoils contain rich mineral deposits. Tropical weather prevails. The hot sun saps human energy and the heat invites debilitating diseases.

Accurate up-to-date demographic data are difficult to procure. Total estimated population exceeds 400 million, with density averaging about 18 per sq. km and ranging from less than 2 in Mauritania to over 90 in Nigeria. Annual population growth averages 2.8%, with over 4% in some areas (e.g., among the Kikuyu in Kenya). Life expectancy averages under 50 years. Wide diversities of language, tradition, and culture exist. Life-styles range from simple tribalistic to urban sophisticate. The populace is predominantly rural, but migration to cities is accelerating. Industrial development has only begun. Weak infrastructures, political instability, complex economic dilemmas, low education, and poor health prevail. Recently, drought and famine have compounded the difficulties.

Obstacles to development: the cluster phenomenon

Numerous obstacles hinder development over the entire region. These problems cluster around key variables which relate to the legacies of colonialism, economic conditions, political considerations, infrastructures, physical and human resources, education, health, urbanization, population growth, and cultural differences. The problem clusters do not exist in isolation. Variables within each cluster interact with variables in other clusters, making difficult the isolation of root causes of slow development. Therefore, our classification of clusters of impediments to development sometimes appears arbitrary. However, the system serves both to demonstrate the complexity of subSaharan development problems and to impose order on this discussion.

THE COLONIAL LEGACY

Although colonialism brought some beneficial change and progress to the region, on balance it left a negative impact. On the positive side, it brought the beginnings of modern transportation and communication systems, increased contacts with the outside world, introduced a civil-service system in which selected Africans participated, expanded external markets for basic raw materials, improved education and medical conditions for some, and promoted law and order according to colonial standards. However, even the positive effects sometimes proved to be a mixed blessing; some even became disastrous for new African nations.

Often colonial inroads began through trade, as expatriate commercial entrepreneurs sought cheap raw materials for their industries. Next, the colonial power established a governate protected by its armed forces, carving out boundaries in accordance with its needs, its desire for empire, or the strength of its military forces. In some cases, the presence of a superior power quieted tribal tensions; in most cases, the new boundary lines realigned African communities into fresh rivalries. In general, the artificial amalgamations have hindered the development of national unity after independence. Often, they spawned political ferment, civil strife, and coups d'état.

Colonialism also fostered economoic dependence. The external trade the colonialists promoted provided them with cheap raw materials for their industries and ready markets for their finished goods. The colonies, in turn, had an export market for their raw goods, increased employment, and access to finished products. Ake (1981) calls such innocent-appearing complementarity fallacious, because (a) much of the subjects' income from trade went into unneeded luxury goods; (b) the colonial powers only developed transport infrastructures adequate for their own needs; and (c) the imported goods flooded African markets with substitutes for traditional items, thus discouraging indigenous craftsmen and encouraging dependence on unskilled labor for primary production. While Ake's views cannot be disregarded, they downplay the employment for many and better entry into industrialization for some which the arrangement brought.

In general, the colonialists encouraged development of a single export commodity, such as peanuts, coffee, or cotton, and discouraged founding of local industry. Klein (1979) relates how, until independence, Senegal had to export peanuts in their shells to France because the latter levied high tariffs to protect its own vegetable-oil industry. Dependence on a single export commodity badly hurt African economies during the recent world recession.

The colonies also became dependent on Western legal and monetary systems. The legal systems imposed overlaid and frequently clashed with indigenous systems of justice. The colonials' monetary systems encouraged dependence on the monetary values of the West.

Colonialism also failed to provide adequate training for development after independence. Most African children received little useful formal education because colonial training was designed for children of the expatriate rulers. Although the colonial overlords did provide some technical training, it proved inadequate for the needs of independence.

Transportation and communication infrastructure

Numerous weaknesses in transportation and communication infrastructures slow subSaharan development. O'Connor (1978) notes that many Africans still transport produce and water on their heads. The colonial powers set up limited transport and communication systems suited to their needs, but insufficient to support effective national development after independence, because of what Ake (1981) calls "disarticulation" or haphazard development of the systems. The railways and roads colonialists built ran parallel toward the sea; they lacked the interconnectedness essential to internal development. I observed a similar situation in Liberia, where multinational corporations built roads and railroads needed to export rubber and iron ore, but insufficient for effective rural development. The rail systems of Zambia, Zaire, Nigeria, Sierra Leone, Mali, Togo, and Senegal all exemplify this phenomenon. In addition, railway gauges often differed between, and sometimes even within, nations.

For their part, the new African nations have failed to give high priority to the improvement of the transport systems they have inherited. Often they lack the financial resources and technical expertise to do so. As a result, the region's transport systems are being developed unevenly. Many roads and bridges remain unimproved. Air transport continues to be costly and inadequate for rapid development. In some cases (e.g., Ethiopia), nations are diverting limited development funds to build a costly new "prestige" seaport.

The region's telecommunication infrastructures rate as the world's poorest. African nations have the world's fewest telephones per unit of population. Internal microwave and telex systems usually connect only principal cities, leaving vast rural areas with few outside contacts. Less than a decade ago, telephone calls were routed through London or Paris. Although this condition no longer prevails, telecommunications remain inadequate for rapid progress.

The region's mass-media systems are poorly developed, unevenly distributed, and as yet too fragile to support speedy development. Many countries of the region still depend heavily on cheap film and program imports. Ugboajah (1985) both demonstrates the need for strong, modern mass-media systems to aid development and deplores the continuing influence of Western values in national media outputs. To these ends, Fisher (1985) calls for more training and cooperation among African nations. The greatest media sophistication centers in urban areas. Distribution systems need improvement; newspapers, and in some cases even radio, do not reach interior rural areas. Such conditions contribute to uneven distribution of information vital to development.

Major weaknesses in the media arise from the lack of well-trained managers, technicians, writers, and producers. This results in an inability to maintain the hardware and to create the indigenous software materials needed for effective education and development. It also perpetuates dependence on the outside world for expertise and materials, and contributes to a widening gap between the region's media systems and more advanced counterparts elsewhere.

POLITICAL PROBLEMS

Shortages of élite political leadership, ethnic divisions, and the lack of overarching political philosophies have led most nations in the region to choose either military rule or one-party government. Both styles perpetuate favoritism, nepotism, and emphasis on the status quo. The number of military states has increased markedly since independence. Some military governments have engaged in political excesses, prolonged internal conflict and even oppression of the people. One-party states, still typical of most nations of the region, essentially depend on a small corps of élite leaders who sustain their control for life. Cronje (1985) observes that sometimes civilian governments are more authoritarian than their military counterparts. Most military and one-party states have failed to achieve their development goals. Diversity of language and culture, inferior education, and lack of economic resources help to account for these failures.

Internal coups d'état are commonplace. For example, since 1982 Nigeria has experienced three, all protests against official corruption, which have contributed to political instability and slowed national progress. Often a coup merely replaces one military regime with another. The presence of outside military forces also has destabilized several nations. Cuban troops have propped up weak puppet regimes in Mozambique, Angola, and Ethiopia. Likewise, South Africa has dominated Namibia.

The political fragility and strife negatively affect development, (a) by forcing leadership to give priority to short-term objectives; (b) by displacing thousands, causing large-scale refugee problems; (c) by diverting urgently needed development resources; and (d) by postponing national unity (World Bank 1983).

Economic problems

Massive economic problems plague the region's development. Most national economies remain underdeveloped, with annual per capita incomes among the world's lowest. Twenty nations in subSaharan Africa are placed within the "world's poorest nations" category, with annual per capita GNP incomes of less than $400 (World Bank 1985). About 70% of the region's people live on the edge of poverty, and World Bank officials consider current economic conditions "grim" and the outlook "bleak" (World Bank 1983). Several subcluster problems help to create this situation.

First, population growth is exploding faster than food production. The *World development report 1983* (World Bank 1983) showed an average annual 2.8% population growth, with food production increasing a mere 1.5%. Most farming remains at subsistence levels. The Food and Agricultural Organization reports that Africans had 12% less home-grown food to eat in 1980 than in 1960. These conditions have led to an 8.4% annual increase in demand for imported foodstuffs which, in turn, drains monetary resources and contributes to unfavorable trade balances. Experts agree that African agriculture is failing to support progress.

Urban economic conditions are faring little better. Heavy rural-to-urban migration of unskilled people leads to high unemployment and poverty in the cities (Sandbrook 1982). Often, the unemployed rural migrants settle in shantytowns where conditions are ideal for fomenting unrest. Urban industrial production is declining (Shaw 1985). Lack of funds, inadequate market bases, and insufficient investment incentives discourage industrial expansion. Most production investments made add little to the basic needs of the poor. These conditions, "labor-rich" and "capital-poor," reflect low incomes for most urbanites, a high dependency on relatives or friends, and an uneven distribution of wealth. Sandbrook (1982) says that the region is not experiencing evolution into more productive enterprises, but "involution," in which more and more share stagnant markets.

The region's nations continue to depend on one or a few export commodities. The World Bank's 1981 *Agenda for action* indicated that over 80% of export income still derives from raw materials; at the same time, internal markets provide little opportunity for import substitution. To date, only a few industries are processing raw materials. Hormats (1983) says the region is caught in a triangular web of economic instability caused by (a) debt-ridden constraints on growth, (b) high unemployment, and (c) pressures from the international trading system. Too-heavy reliance on single export commodities, weak industrial bases, worldwide recession, and declines in demands and prices for African products have placed many of the nations in serious debt. For some, the recent drought–famine cycles only exacerbated the situation, causing what World Bank President A. W. Clausen calls "retrogression rather than progress" (Novicki 1985). The World Bank and the International Monetary Fund were forced to adopt stringent measures for restructuring and repayment of mounting debts.

Other serious economic difficulties exist. Technological dependence continues. In some countries, heavy military expenditure diverts resources for development. Costly internal strife has produced a flood of over five million refugees who have taxed the coffers of several nations. Some nations have spent money on image-making ventures (e.g., big radio and TV operations) in preference to grass-roots projects. In a number of the countries, business and trade remain in the hands of expatriates who export what profits they can. Each drain on the economies of these nations adds to the cluster of economic woes.

EDUCATIONAL PROBLEMS

Educational problems severely hamper subSaharan development. Some arose from the colonial era; others from more recent causes.

Ake (1981) declares that colonial education had two goals: (a) to increase semi-skilled labor and (b) to create a cultural and political atmosphere which would aid colonial dominance. Colonial education did inculcate literacy in Western ideas and did create an African petit-bourgeoisie who became national leaders. Those who attended colonial schools frequently became dissatisfied with working on the land or went abroad for advanced education, often never to return, thus contributing to a "brain drain." The colonials did provide some Africans with education and skills training and this "élite" often assumed leadership after independence. However, a high percentage remained illiterate, a condition which still pertains. In 1984, UNESCO reported that Africa's illiteracy still totalled 150 million, over two-thirds of the productive populace (*Africa News* 1984). The worst illiteracy prevails in the subSaharan nations; for example, literacy rates are still under 5% in Burkino Faso, 10% in Niger, Mali, and Senegal, and 15%

in Ethiopia, Chad, and Sierra Leone. Adedeji (Novicki 1983) projects that the number of African illiterates will grow to 165 million by AD 1990 causing quantitative deficiencies in skills for development.

Prior to the mid-1970s, subSaharan states increased their educational expenditures substantially; however, as Coombs (1985) demonstrates, education's share of national budgets has since declined. African educators face numerous discouragements – overcrowded classrooms, insufficient textbooks, inadequate libraries and equipment, poor teaching, rising costs, and a growing surge of pupils they cannot accommodate. Early school leavers, inadequate teacher training, and the "brain drain" add to their woes.

Coombs (1985) also speaks of the "increasing aspiration/opportunity gap" arising from increased enrollments. National economies are developing too slowly to provide employment for the growing number of graduates; Kenya, for example, can absorb only 22% of secondary and higher education's output.

Shortage of trained technicians causes further grave problems. Many countries still depend on expatriate expertise to install and maintain complicated equipment. Although outside donors (World Bank, USAID, etc.) are supporting crash programs, too few skills-training schemes exist to provide the technicians needed for speedy development.

PHYSICAL RESOURCES PROBLEMS

Several hindrances to development arise from the limited or improper use of physical resources. Only a third of the arable land is being used productively. Subsistence farmers raise food sufficient only for themselves. Repeated cropping (often for export crops such as cotton and tobacco) drains nutrients from the productive soil; high costs discourage replenishment with fertilizers. In tropical zones, farmers employ a "slash and burn" process which provides enough nutrients to raise one crop every several years, but does little to build up soil permanently. In the grasslands areas, as population grows, farmers increase the size of their herds, causing overgrazing which leads to drought and eventual desertification. Many areas face serious water shortages. Water tables have dropped at an alarming rate. Much of the available water is contaminated or brackish; only one African in four has access to clean water (*African Report* 1983).

Although the sea borders half the region's nations, most Africans do not yet regard fish an important source of nutrition and commerce. Meantime, foreign fishing fleets complete with "factory" ships, mostly from the European socialist countries, are taking several million tonnes of fish annually.

Destruction of forests is also accelerating dangerously. Matheson (1983) calls the present wanton exploitation of forests one of the continent's "most pressing problems" and observes that, if the present rate of cutting continues, many countries will have no natural forest left by AD 2000. About

Table 19.1 Population growth in selected African countries.

Country	1985 (millions)	2000 (est.) (millions)
Ethiopia	36.0	54.8
Ghana	14.3	22.9
Kenya	20.2	37.3
Nigeria	91.2	156.5
Sudan	21.8	33.2
Tanzania	21.7	37.3
Uganda	14.7	24.5
Zaire	33.1	52.4

Source: Population Reference Bureau 1985.

90% of the trees cut are being used as cooking fuel; wood is the cheapest source available. The denuding fosters drought cycles and accelerates the development of desert-type ecosystems. Reforestation schemes just started are already insufficient for replacement.

Meantime, certain rich geologic resources are being worked mainly by expatriate corporations because most Africans lack the finance and technical expertise for production and export.

HEALTH AND HUMAN RESOURCES PROBLEMS

Health and human resource issues form yet another cluster of impediments to development. Malnutrition, disease, high infant mortality, and low life expectancy cause debilitating problems. The area has the world's highest child death rate and lowest life expectancy (Shaw 1985). Infant mortality rates per 1000 births are, for example, 200 in Sierra Leone and 193 in Gambia (see Population Reference Bureau 1985 for others). Disease and malnutrition also shorten life spans of the elderly. Tropical killer diseases still take heavy tolls. While vaccination programs are sometimes desultory and clinics and hospitals remain inadequately staffed and equipped, primary health education programs have been launched, medical facilities and treatment are improving, and slowly the battle against disease and malnutrition is being won.

Improved health conditions, coupled with cultural factors (male dominance, importance of large families, security in old age), are contributing to rapid population increase in the region. Population growth is expanding at the rate of 3–4% annually, as Table 19.1 reflects.

The high birth and mortality rates have produced a population structure in which 44.5% are under 15 years old, 52.5% between 15 and 64, and only 3% are 65 or over. The high proportion of dependent youth slows development.

Rapid urban growth accompanies the population explosion. While under

10% of Africans lived in cities in 1960, estimates indicate that one of every three will live there by AD 2000 (*Africa Report* 1983). Many who seek employment in the city fail to find it. Already, governments can no longer supply basic social services. Meantime, the migration drains labor potential from the farms. These dilemmas seem certain to increase.

Other important human problems constrain development. Language, ethnic variety, and cultural differences limit communication and cooperation; the region contains about 800 lingual and ethnic groupings. Some are suspicious of change and cling to tradition. Others feel that they lack a participatory rôle in the society.

THE INTERPLAY OF BARRIER CLUSTERS

Each cluster of development problems in itself represents a significant deterrent to progress. But variables in one grouping interact with those of other clusters to form new impediment combinations. Each new set is unique. For example, the combination of poor transport, failure to industrialize, and high unemployment leads to low per capita incomes and this in turn slows growth. The effects of the clusters discussed earlier and of these new combinations cannot be accurately assessed. However, it appears safe to conclude that the cumulative effects of all hindrances represent a formidable array of barriers to the region's development, and existent conditions seem to bear out this conclusion.

Overcoming obstacles to development

The foregoing discussion outlines the numerous serious obstacles to development faced by the subSaharan nations. Indeed, development progress has regressed during the past decade (World Bank 1983). Shaw (1985) speaks of increasing divergences between and within nations, and warns of the dire implications for satisfaction of basic human needs inherent in current "exponential marginalisation" and of the dangers of extended developmental stagnation.

Despite their good intentions and some positive results, the variety of past and current internal schemes and external projects comprise merely a temporary measure which has failed to grapple with the magnitude and complexities of the region's development dilemmas. The situation calls for extensive and objective re-evaluation of all barriers to progress and of current development policies. The potential of regional cooperative efforts, such as the Economic Community of West African States (ECOWAS), to speed advance should be studied. Where current policies are failing to meet expectations, establishment of new policies sufficiently comprehensive and practicable within the "givens" of the region should be considered.

The nations of the region now face major choices. The World Bank, other external aid sources, and status-quo African élite leaderships have pushed for modernization, renewed incorporation, and dependence on international market forces for progress. The Lagos Plan of Action (ECA 1983) has favored disengagement and greater national and regional self-reliance. Unfortunately, neither has mapped sufficiently comprehensive policies to lift the region to speedy development. Instead, Adedиji (1983) speaks of the region's deteriorating economic conditions and six major crises: (a) chronic food deficits; (b) drought; (c) high import costs; (d) deteriorating trade and mounting balance-of-payment trade deficits; (e) huge external debts; and (f) poor management. The United Nations Economic Commission for Africa is currently seeking to design development strategies which overcome the weaknesses of previous policies. The ECA plans are based on the creation of better political and social environs for development, national self-reliance, and the revival of an African will to advance. Such plans offer hope for designing more comprehensive and effective development policies which place full responsibility on the African nations themselves.

Summary and conclusion

SubSaharan African nations face complex clusters of impediments to their development. Some positive, albeit disparate actions have been taken, but more comprehensive policies which reduce the divergences between nations and provide more productive cooperation are required. Several strategies from renewed incorporation through disengagement to long-term self-reliant schemes have been tried or are being developed.

Overcoming the obstacles to development will require wise utilization of limited resources, sacrifice, willingness to adapt and change, and firm determination on the part of all involved – political leaders, economic planners, outside aid sources, and, above all, individual Africans. The road out of the present stagnation and regression to economic, political, and social development promises to be uphill and arduous. SubSaharan Africans will only be able to ascend it slowly through greater determination, self-reliance, and cooperation among themselves. They – and the outside world – should expect no overnight miracles.

References and further reading

Adedeji, A. 1983. The evolution of the Monrovia Strategy and the Lagos Plan of Action: a regional approach to economic decolonisation. Addis Ababa: Economic Commission for Africa.

Africa News 1984. World literacy rate declining. *Africa News* **23**, 11–12, 14–15.

Africa News 1985. Africa's falling agriculture: battling the odds. *Africa News* **24**,4, 2–7.

Africa Report 1983. Africa: a statistical profile. *Africa Report* **28**,5, 58–60.

Ake, C. 1981. *A political economy of Africa*. Harlow: Longman.
Axon, A. & D. Jamieson (eds.) 1985. *Africa Review, 1985*. Saffron Walden, England: World of Information.
Ayari, C. 1983. What strategy for Africa's development? *African Report* **28**,5, 8–11.
Bissell, R. E. & M. S. Rader 1984. *Africa in the post-decolonization era*. New Brunswick: Transaction Books.
Browne, R. S. 1984. Conditionality: a new form of colonisation? *Africa Report* **29**,5, 14–18.
Coombs, P. H. 1985. *The world crisis in education: the view from the eighties*. New York: Oxford University Press.
Cronje, S. 1985. Military governments in Africa. In *Africa Review, 1985*, A. Axon & D. Jamieson (eds.), 29–30. Saffron Walden, England: World of Information.
ECA 1983. *ECA and Africa's development, 1983–2008*. Addis Ababa: UN Economic Commision for Africa.
Ezenwe, U. 1983. *ECOWAS and the economic integration of West Africa*. New York: St. Martin's Press.
Fisher, H. A. 1985. International cooperation in the development of West African mass media. In *Mass communication, culture and society in West Africa*, F. Ugboajah (ed.), 74–84. Munich: Hans Zell.
Franke, R. W. & B. H. Chasin 1980. *Seeds of famine: ecological destruction and the development dilemma in the West African Sahel*. Montclair, NJ: Allanheld, Osmun.
Ghosh, P. K. (ed.) 1984. *Developing Africa: a modernization perspective*. Westport, Conn.: Greenwood Press.
Gruhn, I. V. 1979. *Regionalism reconsidered: the economic commission for Africa*. Boulder, Col.: Westview Press.
Hormats, R. D. 1983. Africa in the global economy. *Africa Report* **28**,5, 4–7.
Klein, M. A. 1979. Colonial rule and structural change. In *The political economy of underdevelopment: dependence in Senegal*, R. C. O'Brien (ed.), 65–99. Beverly Hills, Cal.: Sage Publications.
Matheson, A. 1983. Africa's fragile environment: the need for a quantum leap to avoid disaster. In *Africa contemporary record: annual survey and documents, 1982–83*, C. Legum (ed.), A180–188. New York: Africana Publishing Company.
Mazzeo, D. (ed.) 1984. *African regional organizations*. Cambridge: Cambridge University Press.
Mtei, E. I. 1984. An evolving relationship. *African Report* **29**,5, 19–21.
Novicki, M. 1983. Adedbayo Adedeji, Executive Secretary, Economic Commission for Africa. *Africa Report* **28**,5, 12–16.
Novicki, M. 1985. A. W. Clausen, President, World Bank. *Africa Report* **30**,3, 14–20.
OAU (Organization for African Unity) 1981. Lagos Plan of Action for Economic Development of Africa 1980–2000. Geneva: Institute of International Labor.
O'Brien, R. C. (ed.) 1975. *The political economy of underdevelopment: dependence in Senegal*. Beverly Hills, Cal.: Sage Publications.
O'Connor, A. M. 1978. *The geography of tropical African development*. Oxford: Pergamon Press.
Population Reference Bureau 1985. *World population data sheet 1985*. Washington DC: Population Reference Bureau.
Sandbrook, R. 1982. *The politics of basic needs: urban aspects of assaulting poverty in Africa*. Toronto: University of Toronto Press.
Shaw, T. M. 1985. *Towards a political economy for Africa*. New York: St. Martin's Press.
Thahane, T. T. 1984. The development challenges ahead. *Africa Report* **29**,5, 48–52.
Ugboajah, F. O. (ed.) 1985. *Mass communication, culture and society in West Africa*. Munich: Hans Zell Publishers.

World Bank 1981. *Accelerated development in subSaharan Africa: an agenda for action*. Washington, DC: International Bank for Reconstruction and Development/World Bank.

World Bank 1983. *World development report 1983*. Washington, DC: World Bank/Oxford University Press.

World Bank 1984. *Population change and economic development*. New York: Oxford University Press.

World Bank 1985. *World development report 1985*. New York: Oxford University Press.

20 *World (Third?) within a world (First): Canadian Indian people*

EVELYN PETERS

Attempts to develop indexes of development and underdevelopment are complicated by the lack of homogeneity within countries. Canada is a case in point: the demographic characteristics and circumstances of daily life for Canadian Indian people parallel those of people in many underdeveloped nations. Differences in the social and economic conditions of Indians and non-Indians almost create the impression that Canadians who are Indians live in a different world from Canadians who are not.

Researchers have used a variety of models and definitions of development to interpret the situation of Indian people in Canada, and Lithman (1983) has written a good critical review of this work. The primary purpose here is not to develop alternative models of the processes involved. Instead this paper raises a number of questions about the identification of First, Third, and perhaps Fourth Worlds engendered by the situation of Indian people in Canada.

The first section of the analysis uses some common demographic indicators to compare Indians, Canadians, and other countries. Then we turn to the socioeconomic conditions of Indians and other Canadians. Finally we concern ourselves briefly with questions raised in recent years by Canadian Indians, along with other aboriginal peoples internationally, about aspects of the situation of aboriginal peoples which differentiate them from Third World nations. But first, a note on the definitions and sources used in this paper.

Definitions and sources

There are a number of difficulties involved in the identification of the Indian population of Canada for statistical comparisons. The 1981 Special Census of the Native Peoples used three main categories: Status Indians, Non-Status Indians, and Mêtis. In that year the federal government only recognized a legal obligation to Status Indians as defined under the 1951 Indian Act, and because the definitions of Mêtis and Non-Status Indian varied widely, statistics for other than Status Indians were not consistently collected by most service agencies and government departments. Where they were collected they were frequently not comparable. The limited information available suggests that in many ways the situation of Mêtis and Non-Status Indians parallels that of Status Indians. Data limitations however lead us to concentrate in this paper on the last of these groups.

Statistics from the Department of Indian and Northern Affairs and the 1981 Canadian Census form the bulk of the data used in this paper. With its question on ethnic origin and its special tabulations for Indian people, the 1981 Census for the first time provided detailed statistical information about the current demographic and socioeconomic characteristics of Canada's Indian population. The Census definition of Status Indian was based on individuals' self-identification of origin. Although the terms "Status" and "Registered" Indian are frequently used interchangeably, the Census Status Indian population may be slightly different in composition from the Department's Registered Indian population.

The Department of Indian and Northern Affairs defined Indian status according to the terms and conditions of the Indian Act 1951, rather than by individuals' self-identification. The Department's 1981 statistics for Registered Indians excluded the Mic Mac Indians, who were negotiating their status with federal and provincial governments, and Status Indian women married to Non-Indian men. Individuals from both of these groups may have identified themselves for Census purposes as Status Indians. Non-Indian women married to Status Indian men were also counted as Registered Indians although they were not of Indian origin.

Approximately 3.3% of the 1981 Status Indian population listed in the Census were not recognized as Registered by Indian and Northern Affairs Canada, and a slightly smaller number of individuals who were Non-Indian by origin were Registered Indians by marriage. The populations involved were probably small enough for there to be no significant effect on the basic measures used in this paper, and any effect should make the Registered Indian population more like the total Canadian population. For the reader's information, however, in the analysis which follows the term "Registered Indian" refers to the data set used by Indian and Northern Affairs Canada, while "Status Indian" refers to the population described in the 1981 Canadian Census.

A final cautionary note: as anyone working with cross-cultural and international statistics must be aware, comparability is complicated by problems such as variations in national statistical practices, different standards of record-keeping, cultural differences in definitions, and perhaps most importantly variations in the meaning and significance of events and circumstances in different cultural milieus. The data should thus be construed only as indicating trends and characterizing major differences between nations. Rather than supplying a complete description in the text of the data

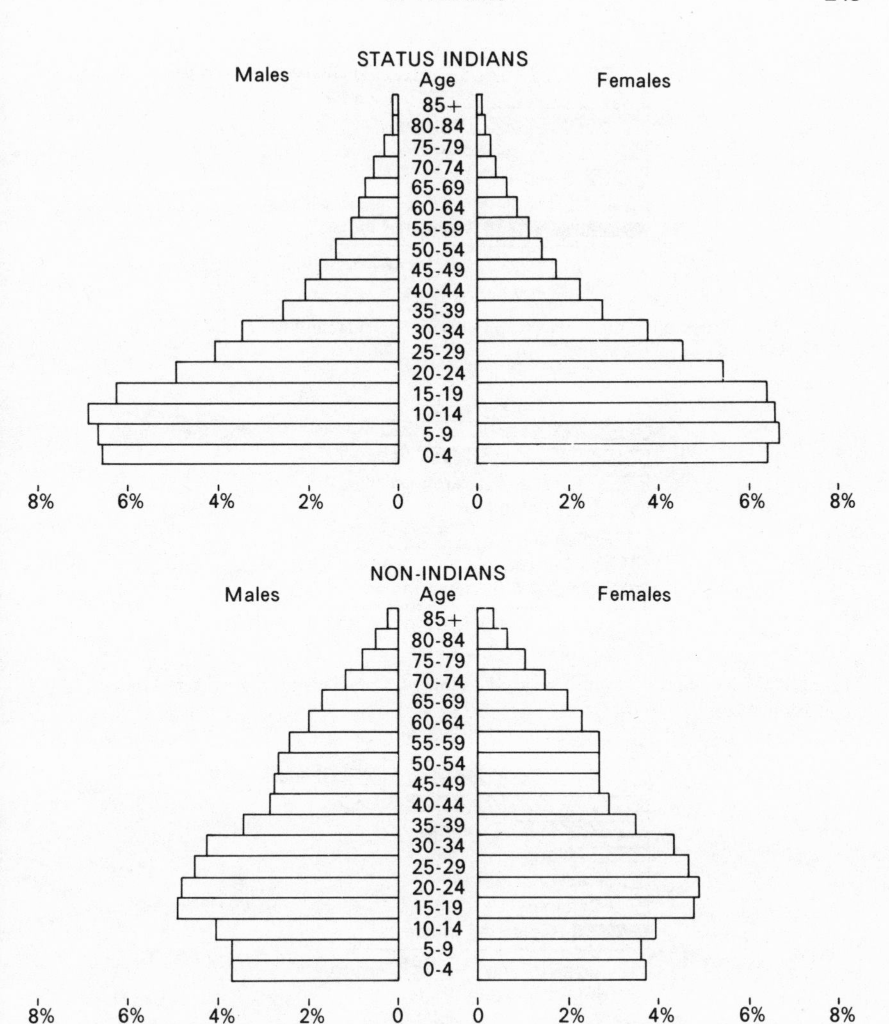

Figure 20.1 Age–sex profile of Status Indians and the non-Indian population, Canada, 1981.

Source: 1981 Census of Canada.

limitations for each of the measures used here, the reader is referred to the data sources where such information is presented.

Demographic indicators: national and international comparisons

The age–sex structure of the Status Indian population differs considerably from that of the Non-Indian population, and exhibits to a large degree the

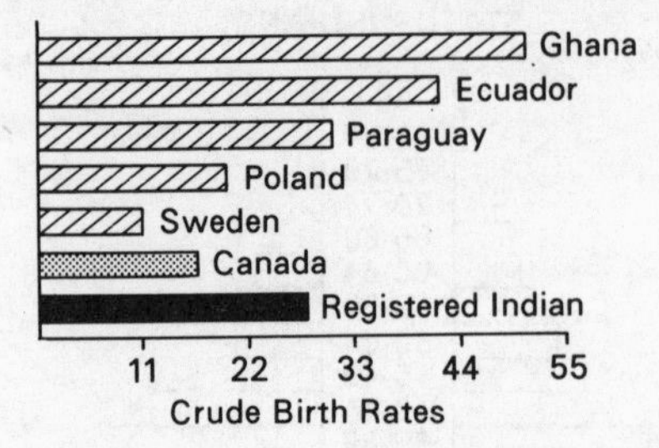

Figure 20.2 Crude birth rates: selected countries, Canada, and the registered Indian population, 1981.

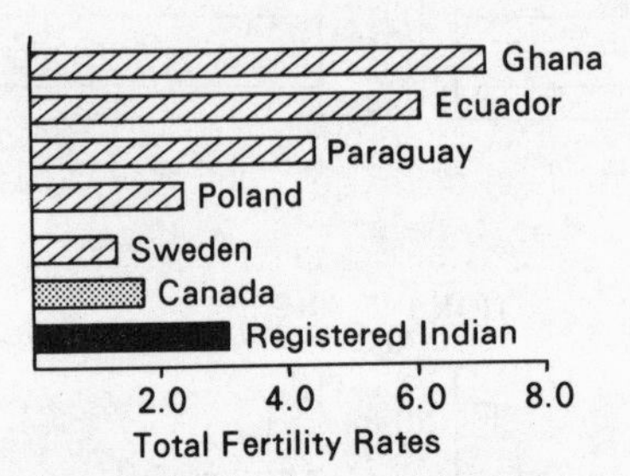

Figure 20.3 Total fertility rates: selected countries, Canada, and the registered Indian population, 1981.

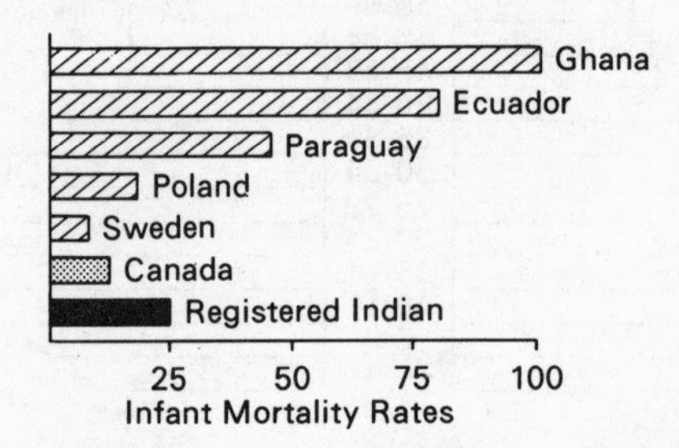

Figure 20.4 Infant mortality rates: selected countries, Canada, and the registered Indian population, 1981.

Sources: UNICEF 1984. *The state of the world's children*. Oxford: Oxford University Press. 1984 Statistics Canada, *Vital Statistics*, Cat. Nos. 84–204, 84–206.
Indian and Northern Affairs Canada 1985 *Population projections of registered Indians, 1982–1996*. Ottawa.

broad base and narrow tip characteristic of Third World countries with high birth rates and declining but still relatively high death rates (Fig. 20.1). Like most Third World countries, a large proportion of the Status Indian population (approximately 43%) is composed of children of 0–14 years of age.

Although birth rates have declined in recent years, as the base of the pyramid indicates, 1981 Indian birth and fertility rates were considerably higher than those found in First World countries. Crude birth rates for 1981 presented in a UNICEF list of 130 countries with a population of three-quarters of a million or more display a range from 55 annual births per 1,000 population for Ghana to 11 annual births per 1,000 population for Sweden (Fig. 20.2). The Registered Indian birth rate falls near the middle of the range, along with most South and Latin American countries. Canada, with a crude birth rate of 15.3, is like other countries in the First World.

If a large proportion of a population's women are in their childbearing years, as is the case among Indians, crude birth rates may disguise patterns of fertility. Total fertility rates that sum age-specific fertility rates represent the number of children on average each woman has while passing through her child-bearing years. In 1981 total fertility rates for Registered Indians were almost double those of the Canadian population, indiating once more the difference between these two populations (Fig. 20.3).

Infant mortality rates have often been cited as a sensitive indicator of the socioeconomic conditions of a population. A comparison of infant mortality rates in 1981 again suggests that Indians and other Canadians occupy different "worlds." Although Registered Indian infant mortality rates can be grouped with those of other First World countries in the UNICEF classification for countries with 25 or fewer infant deaths per 1,000 live births annually, Indian infant mortality rates are nevertheless twice as high as those in the total Canadian population (Fig. 20.4).

Indians and other Canadians: the persistent gap

In 1947–8 a Special Joint Committee of the Senate and House of Commons was appointed to re-examine the Indian Act. In the course of its deliberations the committee heard shocking evidence concerning the depressed social and economic conditions on Indian reserves. The Committee produced a series of sweeping recommendations aimed at bringing the social and economic circumstances of Indian people into parity with the rest of the Canadian population.

Since that time there have been dramatic increases in expenditures on and services for Indian people. In absolute terms it is undeniable that the circumstances of Indian peoples have improved greatly. Infant mortality rates for example declined from 76.3 annual infant deaths per 1,000 population in 1961, to 21.8 per 1,000 in 1981.

A comparison of changes for Indians with changes occurring for *all* Canadians over the past few decades presents a slightly different view. Using infant mortality rates available on a yearly basis for Registered Indian and the total Canadian population it becomes evident that although Indian infant mortality rates have decreased substantially, they are still more than twice as

Table 20.1 Infant mortality rates for the registered Indian and total Canadian populations, 1961–81.

Year	Canadian	Registered Indian	Ratio: Indian/Canadian
1961	27.2	76.3	2.8
1962	27.6	74.9	2.7
1963	26.3	70.4	2.7
1964	24.7	63.9	2.6
1965	23.6	52.6	2.3
1966	23.1	52.4	2.3
1967	22.0	53.6	2.4
1968	20.8	48.6	2.3
1969	19.3	41.1	2.1
1970	18.8	*	*
1971	17.5	45.2	2.6
1972	17.1	47.5	2.8
1973	15.5	40.8	2.6
1974	15.0	39.3	2.6
1975	14.3	38.6	2.7
1976	13.5	32.1	2.4
1977	12.4	33.5	2.7
1978	12.0	26.5	2.2
1979	10.9	28.2	2.6
1980	10.4	24.4	2.3
1981	9.6	21.8	2.3

* Data not available.

Sources: Health and Welfare Canada, Medical Services Branch 1972–81. *Annual reports*. Health and Welfare Canada, Medical Services Branch 1978. *Health data book*. Statistics Canada. *Vital statistics*. Catalogue Nos. 84–201, 84–204, 84–206.

great as those of the total Canadian population, and have been more than twice as great for twenty years (Table 20.1).

Perhaps the important question arising from a discussion of changes in the situation of Canadian Indians over time is not the degree of relative improvement, but why the gap between Indians and Canadians persists. The 1981 Census shows significant differences in rates of employment, levels of income and education, and housing conditions (Table 20.2). Approximately twice as many Indian as Non-Indian adults have less than Grade 9 education, and a much smaller proportion of the Status Indian population participates in the labor force. A comparison of the proportion of Indians and Non-Indians employed shows a similar pattern. Indians who are working have more difficulty finding full-time employment, and housing conditions are worse and common amenities fewer for Status Indians than for other Canadians.

The continuing existence of these considerable differences in standards of living raises questions about the nature of the development process within, as well as between, countries. Although the situation of Indian peoples may

Table 20.2 Education, employment, income, and housing among status Indians and non-Indians, Canada 1981.

Variable	Non-natives	Status Indians
Education		
Population 15 and over not attending school, with less than Grade 9 education (%)	22.0	41.4
Employment		
Population 15 and over not in the labor force (%)	35.0	53.8
Males 15 and over employed (%)	74.4	51.3
Those who worked in 1980, who worked 40–52 weeks, mostly full-time (%)	41.8	24.9
Income		
Average individual income, population 15 and over ($)	13,100	7,780
Housing		
Houses that need major repairs (%)	6.5	19.5
Houses lacking bathrooms (%)	1.1	21.1

Source: 1981 Census of Canada.

have improved in recent decades their continuing position as the poorest and most economically marginal sector of Canadian society suggests that serious examination is needed of the factors that perpetuate the existence of these two "worlds" within a First World nation.

Canadian Indians: a Fourth World?

The first use of "Fourth World" to describe Canadian Indian peoples appears in the work of George Manuel and Michael Posluns (1974). In recent years the term has been used more widely to denote the encapsulated or enclaved societies of many aboriginal peoples internationally. While a simple definition of what constitutes the "Fourth World" remains elusive, aboriginal peoples often reject standard political and economic analyses of traditional development models applied to the other three worlds. Aboriginal peoples' relationship with the dominant society involves a fundamental challenge to their ability to order their lives by their own values and understanding.

The challenge to indigenous culture is not unique to aboriginal societies: it has occurred in all colonized nations. Among the aboriginal peoples it could be argued however that the assault has been more pervasive and all-encompassing. In the case of Canadian Indians until recently, very few of the decisions about the context and direction of daily life lay outside the purview of the Department of Indian Affairs.

Stea & Wisner (1984) argue that the continuing encroachment on what has traditionally been Indians' land represents an indirect assault on Indian culture. They write that, given Indian people's cultural attachment to the land, the continuing erosion of their land base constitutes a fundamental attack not only on the organization of their daily life but also on their religion and on Indians' cultural understanding and interpretation of their lifeworld.

The case of law and justice illustrates another assault on Indian culture. In 1981 Indians made up 5.9% of the inmate population in federal penitentiaries, although they comprised only 1.5% of the Canadian population. Menno Wiebe (1984, p. 17) argues that the over-representation of Indian people in the criminal system occurs not because Indian people are somehow "lagging behind in the matter of legal obedience." Instead, he says, a major part of the problem lies in the imposition of European law which is frequently incongruent with Indian values, and law-keeping systems which are alien to Indian ways of administering justice.

A more direct challenge has involved the abrogation of Indians' ability to teach their cultural heritage to their children. The outlawing of Indian religion and Indian ceremonies in the last century, and until the 1960s the removal of young Indian children from their parents to distant residential schools where they were forbidden to speak their native language, constitute two aspects of the attack on Indian culture. The wide-scale removal of Indian children from their families and their placement in the care of child welfare authorities has taken place more recently. Patrick Johnston's (1983) study on Indian children and the child welfare system indicates that between 1955 and 1964 the represntation of Indian children in care in British Columbia increased from less than 1% to 34.2% of all children in care. This pattern was being repeated in other parts of Canada. In 1980 Status Indian children were represented in the child welfare system at approximately four-and-one-half times the rate for all children in Canada (Johnston 1983, p. 57). By the 1980s some reserves had lost almost a generation of their children as a result.

The attack on culture has been costly in human terms for Indian people. Anastasia Shkilnyk's (1985) study of the Indians at Grassy Narrows illustrates the effect on native individuals and communities. The events on which Shkilnyk focuses occurred between 1960 and 1980. First, the tribe was relocated to a new reserve where their ability to engage in traditional subsistence activities was severely limited. Second, Non-Indian patterns of space and order were imposed in the plan of the new settlement, disrupting clan and kinship organization in daily life. Mercury poisoning in the water system came last, with its implications for health and ways of making a living. These changes came quickly and the Indian community had no control over the direction or rate of change, and no explanation of its implications. The people, Shkilnyk argues, found it impossible to create alternative ways of living and understanding in the face of such total change, and the result was the disorganization and destruction of the community. Shkilnyk (1985, p. 17) links high rates of alcoholism, accidental deaths

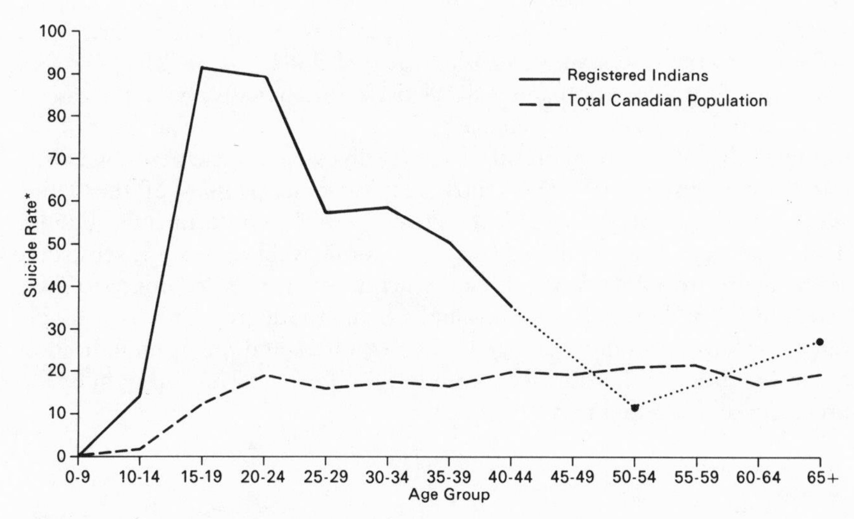

Figure 20.5 Deaths by suicide for registered Indians and the total Canadian population, 1981.

* Annual number of deaths per 100,000 population.
Sources: Health and Welfare Canada, Medical Services Branch 1982. *Annual review*. Ottawa. Statistics Canada, 1983. *Vital statistics*, Cat. no. 84–206. Ottawa.

frequently related to alcohol consumption, and suicide especially among the young people, to "depression, hopelessness, a loss of moorings, an erosion of the symbols and points of reference essential to life's continuity."

While the Grassy Narrows case is in some ways unique, many Indians in Canada have experienced a similar loss of control over the direction of their lives and a similar separation from a sense of its meaning. Rates of accidental and violent death in the Indian population generally are very high. In 1978 "accident, violence, or poisoning" was listed as the primary cause of death for approximately 5% of the Canadian population; it was the principal cause of death for 35% of the Registered Indian population. Suicide rates among Registered Indians, especially among the young, tell a tale similar to the Grassy Narrows story, with suicide rates for Indians between the ages of 15 and 24 approximately six times the rates for that age group nationally (Fig. 20.5).

A number of Indian organizations see increasing Indian self-determination as the first step in ameliorating continuing social problems among Indian communities. Sol Sanderson, president of the Federation of Saskatchewan Indians in 1980, maintains that many of the problems experienced in Indian communities are symptoms of Indians' lack of effective control over virtually every aspect of their lives. He writes: "In Saskatchewan, Indians have decided to confront this problem by taking control over our communities and our lives" (Opekokew 1980, p. 154). Contemporary Indian leaders and communities are exploring the possibility and structure of Indian self-government in an attempt to make it possible to order their lives according to their own values (Little Bear *et al.* 1984).

It is not clear what comparable data exist internationally, but if the case of Indian people in Canada is indicative, researchers must take account of the elements that structure aboriginal peoples' lives. These elements and their effects may differentiate aboriginal peoples from peoples of the Third World. Perhaps measures could be created which would include, besides demographic indicators and indexes of economic wellbeing, some sense first of the degree to which a people can structure their lives according to their own cultural understandings and values, second the degree to which a people can pass on their cultural heritage to their children, and finally an indication of the disruption in the fabric of community and personal life wrought by the processes of "development."

Conclusion

The process of development and the creation of First and Third Worlds has most frequently been seen as a process occurring between nations. The case of Indian people in Canada indicates that there must be a disaggregation to take account of relations between different populations within nations as well. At the same time, however, it is suggested that models applicable to the Third World cannot be simply extended to the problem of aboriginal peoples: the uniqueness of their circumstances, in this instance of the circumstances of Indians in Canada, must be taken into account.

References

Johnston, P. 1983. *Native children and the child welfare system*. Ottawa: The Canadian Council on Social Development.

Lithman, Y. G. 1983. *The practice of underdevelopment and the theory of development: the Canadian Indian case*. Stockholm: Studies in Social Anthropology, University of Stockholm.

Little Bear, L., M. Boldt & J. A. Long 1984. *Pathways to self-determination: Canadian Indians and the Canadian State*. Toronto: University of Toronto Press.

Manuel, G. & M. Posluns 1974. *The Fourth World: an Indian reality*. Don Mills: Collier Macmillan Canada.

Opekokew, D. 1980. *The First Nations: Indian Government and the Canadian Confederation*, Saskatoon: Federation of Saskatchewan Indians.

Shkilnyk, A. M. 1985. *A poison stronger than love: the destruction of an Ojibwa community*. New Haven: Yale University Press.

Stea, D. & B. Wisner 1984. Introduction: the Fourth World, a geography of indigenous struggles. *Antipode* **16**,2, 3–12.

Wiebe, M. 1984. *Native culture and Canadian law: a cultural look at native people and the Canadian justice system*. Kingston: Queen's Theological College.

21 *The United States as a Third World country: race, education, culture, and the quality of American life*

JOEL LIESKE

American scholars are generally not accustomed to viewing the United States as a Third World country. Neither do many seem favorably disposed toward racial–ethnic and cultural interpretations of differences in the quality of American life. Nonetheless, it is evident that many differences that allegedly separate the US and the Third World may be more pronounced in degree than type (Sharkansky 1975). In addition, it is clear that on a number of social indicators, the quality of life in some American cities actually falls *below* the standards that prevail in many major cities of the Third World.

In terms of economic wellbeing, the United States, with a per capita income of $11,347 in 1980, led most developed nations ($8,477) and the overwhelming bulk of underdeveloped nations ($794), with the notable exception of several oil-rich sheikhdoms in the Middle East (Sivard 1983, pp. 29–31). Yet, the infant mortality rate (number of deaths per 1,000 live births) is higher in Washington, DC (27.3), Baltimore (23.2), St. Louis (22.5), Detroit (22.3), and Chicago (21.3) than it is in such Third World incubators as Havana (17.5), Bangkok (17.2), Beijing (14.8), Bogata (11.5), Cairo (11.0), Singapore (10.8), and Hong Kong (9.9). On the other hand, the worst American cities do not match the depressing infant mortality levels reported in the Third World cities of Manila (69.5), São Paulo (56.1), Rio de Janeiro (51.5), Delhi (50.2), Mexico City (45.2), and Buenos Aires (27.5) (Marlin *et al.* 1986, pp. 582–3). Curiously, the mixed performance of American and Third World cities can not be wholly explained by differences in the availability of medical care. For instance, Washington DC, with one doctor for every 205 residents, has a better physician : patient ratio than Beijing (235), Havana (336), Madrid (356), Cairo (630), Bangkok (1,256), and Hong Kong (1,265), but also, as noted above, a significantly worse infant mortality rate.

And though the United States is typically viewed as a civilized nation, cross-national crime statistics reveal that many American cities are quite violent and uncivilized. Thus the world cities that rank highest in the number of homicides per 100,000 population are all located in the United States. Based on data reported by Marlin *et al.* (1986; p. 603), they are respectively St. Louis (58.5), Detroit (41.7), Washington, DC (35.0), Dallas (33.3), Los Angeles (29.7), Chicago (29.2), Baltimore (29.0), New York (25.8), and Philadelphia (21.5). All have higher homicide rates than such Third World cities as Rio de Janeiro (19.8), São Paulo (17.7), Bangkok (12.5), Delhi (3.7), and Bombay (3.0). Similar reported statistics on the total number of criminal offenses per 100,000 population show that the urban crime problem is far worse in Boston (14.1), St. Louis (13.8), Dallas (12.3), Detroit (11.9), Washington, DC (10.7), San Francisco (10.6), Los Angeles (10.3), and New York (10.3) than it is in Addis Ababa (3.4), Seoul (2.0), Hong Kong (1.7), Santiago (1.6), Singapore (1.6), Bangkok (1.2), Buenos Aires (0.9), Manila (0.8), Delhi (0.7), Bombay (0.3), Rio de Janeiro (0.3), São Paulo (0.3), Bogata (0.2), and Jakarta (0.1) (Marlin *et al.* 1986, pp. 602–3).

Even the vaunted American system of education does not always provide the level of individualized instruction provided in many Third World cities. Since 1900 educational studies in the US have identified the student : teacher ratio as one of the best predictors of success in standardized achievement tests (Boyer & Savageau 1985, p. 220). While the cities with the worst student : teacher ratios tend to be located in the Third World, some Third World population centers such as Shanghai (11.2), Buenos Aires (11.8), and Beijing (13.4) rank in the top five, surpassing even Boston (15.4), St. Louis (16.7), and Philadelphia (16.9) (Marlin *et al.* 1986, p. 597).

A final point of comparison concerns several conditions the US shares with the Third World. First is the condition that developmental theorists have labeled as "dualism" – the existence of economically advanced and backward subcultures in close promixity to one another. Though this condition is almost universal, it tends to be more pronounced in underdeveloped than developed countries. Despite past civil rights gains in the United States, the currents of dualism continue to run strong and deep throughout the structure of American society. One cause of concern is the growing divergence between average black and white family income.[1] Another is the increasing inequality in the distribution of national income since 1960 (Pear 1982). Yet another is the expanding economic disparities between declining central cities and their affluent outlying suburbs. The postwar flight of business, industry, and the middle and upper-middle classes, which peaked during the 1970s, may now be abating. But in its wake, it has left a large and growing, perhaps permanent underclass (Auletta 1981). Composed largely of the nation's poor, unemployed, dispossessed, and unassimilated racial minorities, this underclass has given many cities the appearance, if not the reality, of social dumping grounds.

A second condition is the uneven economic, social, and political development of different metropolitan areas and regions. Historically the South has been the most backward region in the US. Notwithstanding regional economic convergence (Weinstein & Firestine 1978) and the rise of the sunbelt cities (Perry & Watkins 1977), it still remains an underdeveloped region. Since 1929 per capita income in the South has steadily increased from 53% to 84% of the national average (Weinstein & Firestine 1978, p. 50). But even today, many businesses and industries are attracted by its "favorable" business climate: low pay scales, state right-to-work laws, weak unions, a generally docile labor force, and low taxes.[2]

A third condition is cultural pluralism. Even before the Immigration Act amendments of 1965, the US was populated by many different racial and ethnic groups. But since it opened its portals to immigrants with non-European backgrounds, even greater changes have occurred in the composition and character of American society. For instance, from 1930 to 1960 about 80% of US immigrants came from European countries or Canada. From 1977 to 1979, however, only 16% did, while Asia and Latin America accounted for about 40% each. By 1979 the nine leading source countries for legal immigration were Third World nations; namely, Mexico, the Philippines, Korea, China and Taiwan, Vietnam, India, Jamaica, the Dominican Republic, and Cuba. In tenth place with only 3% of the total was Great Britain. Based on immigration trends reported by Fallows (1983, p. 46), the legal foreign-born comprise about 18 million or almost 8% of the US population. In addition, there are an estimated 6–9 million illegal immigrants. To these figures must be added an average of 600,000 legal and at least 500,000 illegal immigrants each year. With many, if not most, settling in large central cities, the traditional haven of refuge and sanctuary for newly arrived groups,[3] the nation's metropolitan areas are once again peopled by every race, skin color, and social condition.

As a consequence, American cities and metropolitan areas provide systematic comparisons and striking contrasts in the conditions of life of a kind not found inside many industrial countries. Moreover, quantitative data for them are generally richer, more reliable, and more comparable than for Third World societies. The American laboratory thus provides a unique opportunity for scholars to study rigorously a variety of Third World conditions within a comparative context.

This study will assess the effects of three conditions – racial dualism, economic development, and political culture – on the quality of American life. This will be done by analyzing comparative data for 243 US metropolitan areas. My objectives are two-fold. One is to develop and test an empirical theory of life-quality differences. The second is better to understand the conditions that affect urban and human development.

A theoretical model

A review of the literature suggests that many life-quality differences may be explained by three broad, theoretical perspectives: (a) racial dualism, (b)

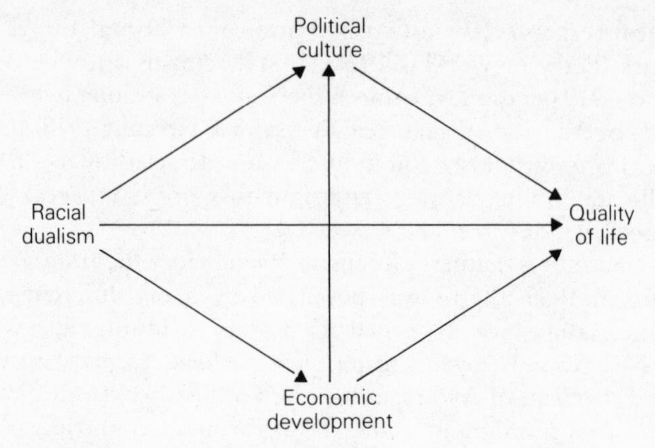

Figure 21.1 A quality-of-life model.

economic development, and (c) political culture. The first emphasizes the pervasive differences in social wellbeing between whites and racial minorities who live in close proximity. The second, by comparison, stresses a number of conditions that are associated with modernity; namely urbanization, industrialization, wealth, and education. The third emphasizes the common beliefs and behavior that differentiate one ethno-religious or regional grouping of people from another.

The hypothesized joint effects of racial dualism, economic development, and political culture of the quality of life are diagrammed in Figure 21.1. Since the black and Hispanic communities are generally considered to constitute separate national subcultures and since they are generally viewed as less developed economically, both the developmental and cultural factors are assumed to vary as a function of racial dualism. The direct paths from racial dualism and economic development to political culture are premised on the assumption that some cultural differences may be due to differences in racial dualism and modernity. Thus, in the face of controls for dualistic and developmental variables, the cultural effects which are computed under this model must be regarded as conservative. The purpose of this section is to develop the key assumptions that undergird the theories of racial dualism, economic development, and political culture.

RACIAL DUALISM

The theory of racial dualism seems basic to a number of studies which have emphasized the continuing racial legacy of America's colonial past. Thus, race has been identified as a critical factor in explaining: (a) the social and political development of the South (Key 1949, Bartley & Graham 1975, Gastil 1975), (b) income inequality among whites (Reich 1977), (c) the dual labor market (Doeringer & Piore 1975), and (d) cultural and regional variations in reported crime (Pettigrew & Spier 1962, Gastil 1975).

The hypothesized effects of racial (and social) dualism on the quality of life are based on several assumptions. First, dualism theory assumes that minority concentration (black and Hispanic) is a key differentiating characteristic both within and across US metropolitan areas. Second, since most quality-of-life indicators represent averages over social aggregates, it is further assumed that reduced standards of wellbeing for blacks and other racial minorities will be reflected in lower overall statistics. Finally, it is assumed that racial discrimination not only hurts blacks but also tends to retard societal progress for whites as well (Reich 1977).

ECONOMIC DEVELOPMENT

This theory has many facets, but associated with most variations are usually the notions of economic productivity, a diversified economy, wealth, education, and material consumption. All are generally assumed to improve the human condition and hence the quality of life. In addition, the theory usually assumes that economic development is a necessary prerequisite for social and political development. This assumption is based on studies that have linked indicators of economic development with social wellbeing (Smith 1973), democratic political cultures (Almond & Verba 1963, Verba & Nie 1967), political trust and support of government (Erikson *et al.* 1980, p. 98), and the quality and quantity of government services (Sharkansky 1970).

Despite its ability to predict many life-quality indicators, development theory suffers from several limitations. First, the elements of economic development – urbanization, industrialization, wealth, and education – do not always occur together or in a causal linear sequence, as was once believed (Winham 1970). For example, the state of Iowa, which is perhaps the most agricultural, has consistently scored the lowest on illiteracy (Gastil 1975, p. 118). Second, the impact of economic development is not always beneficial. For instance, there appears to be a trade-off between economic vitality and environmental quality in large and medium-sized industrial cities (Liu 1976). Third, some presumed indicators of development, such as education, may actually reflect cultural differences – witness for example Utah, a state that ranks highest in the number of college graduates per capita, but fifth lowest in per capita income.

POLITICAL CULTURE

Several assumptions appear to undergird a cultural interpretation of life quality differences. One is a rejection of environmental determinism. Thus culture theory (Gastil 1975, p. 26) assumes "that different people make different uses of the same environment and that people use material goods rather than the other way around." As Gastil (1975, p. 26) argues:

> . . . The Great Plains was one geographical fact that was populated by a variety of different people. In the South it has been populated by

Southern Baptist farmers, highly individualistic and inclined to violence. At the northern end the settlers were in large measure descendants of New Englanders and Scandinavians with a formal Lutheranism and political-economic ideals of the cooperative movement. Further east, the New Englanders and Germans of Wisconsin built up a progressive, serious, educationally-oriented society that contrasts markedly with the old American society of central Illinois and Indiana.

A second assumption is derived from Zelinsky's (1973) "Doctrine of First Effective Settlement." This doctrine assumes (Gastil 1975, p. 26) that cultural differences are primarily due to "variations in the cultures of the peoples that dominated the first settlement and the cultural traits developed by these people in the formative period (where they are significant)." This assumption does not deny the possibility of later change, the overwhelming of the original cultural imprint by a later group, or the countervailing influence of a competing élite. However, later cultural streams are generally considered to exert secondary effects which are modified by the original streams (Elazar 1970, 1972).

Perhaps the most compelling theory of "first effective settlement" patterns in the US is Elazar's (1970, 1972). In his view, American political culture is the product of three native cultural streams and the political-cultural leanings of various ethnic and religious immigrant groups. On the basis of the historical migration patterns of the native streams and their affiliated ethnic and religious groups, Elazar has produced a geopolitical mapping of the US that permits classification of states, metropolitan areas, and rural areas into combinations of three competing subcultures. Taking into account the central characteristics that govern each subculture and their respective centers of origin, Elazar has designated the three subcultures as "individualistic," "moralistic," and "traditionalistic."

According to Elazar, the "individualistic" culture originated in the Middle Atlantic states and was powerfully shaped by free-market values, ethnic and religious pluralism, and partisan politics. As a consequence, individualistic cultures supposedly emphasize the centrality of private concerns, restrict government action to predominantly economic activities, and favor professional, job-oriented over amateur, issue-oriented politics. The cultural seat of Elazar's "moralistic" culture, by comparison, is Puritan New England, where individual enterprise was harnessed by dominant communal norms that stressed religious orthodoxy, social and moral improvement, and democratic self-government. Consequently, moralistic cultures allegedly emphasize a commonwealth vision of government, moral politics, government intervention on behalf of the public welfare, and the virtues of widespread democratic participation. Finally, Elazar locates the core of the "traditionalistic" culture in the southern states where slavery and a plantation economy nurtured a radically individualistic social system and a

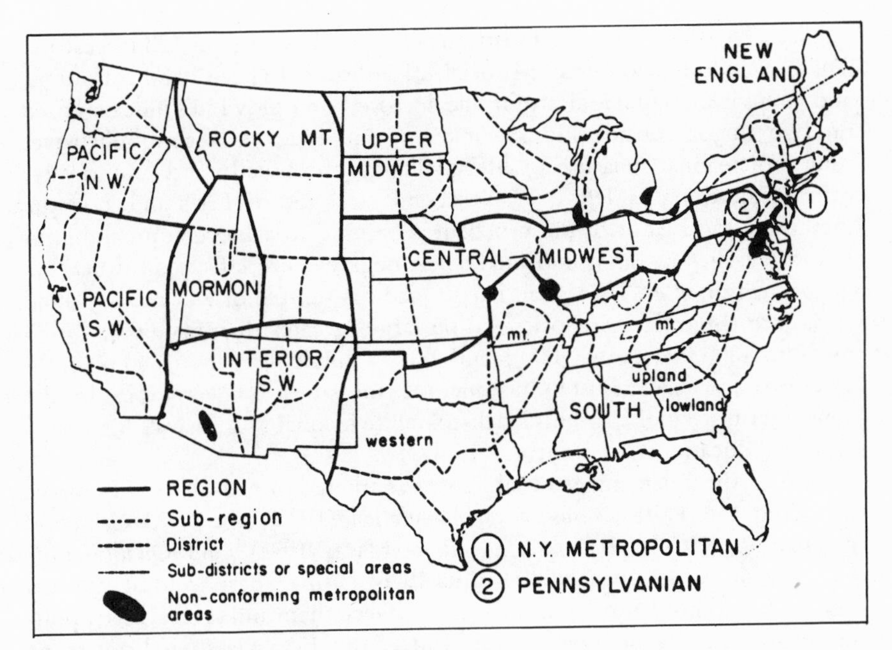

Figure 21.2 Cultural regions of the United States.

gentry-dominated political order. As a result, traditionalistic cultures are seen to stress the values of social order, instititutional hierarchy, élite control and deferential politics.

A third cultural assumption concerns the hypothesis of regional homogeneity. According to Gastil (1975), the US may be divided into geographically contiguous areas that are relatively homogeneous with respect to such cultural factors as history, national origin, religion, dialect, regional consciousness, social norms, and political standards. Based on these factors he has partitioned the US into 13 distinct cultural regions. See Figure 21.2.

A final assumption is that the quality of life in any metropolitan area or region ultimately depends on the characteristics of the people who live there; namely, their values, beliefs, education, job skills, and behavior – in short, their way of life (Lieske 1984).

Methodology

Data on the quality of American life and selected indicators of racial dualism, economic development, and political culture were gathered for 243 metropolitan areas designated in the 1970 US census. This sample was chosen in order to use Ben-Chieh Liu's (1976) quality-of-life indicators. The dependent variable employed in this study is Liu's (1976) overall index of life quality. The Liu index is a statistically-weighted average of 123 indicators. These include 18 economic, 54 social, 21 political, 13 health and education, and 17 environmental indicators. The decision to employ Liu's index is based on two major considerations. First, the Liu index is a comprehensive, multidimensional measure of life quality. Conceptually and empirically, little if anything is left out. A second strength of Liu's index is the methodological rigor of the procedures he used to achieve comparability. Because earlier research indicated some negative size effects on urban life quality (Elgin *et al.* 1974), Liu partitioned his sample of 243 metropolitan areas into three size categories: (a) 65 large (greater than 500,000), (b) 83 medium (200,000 to 500,000), and (c) 95 small (less than 200,000). In addition, Liu has adjusted his income measures for variations in the consumer price index and his black-to-white inequality measures for differences in educational attainment.

Data on the independent variables were gathered from a variety of sources including the 1970 *Census of the population* (1973), the 1972 *Census of governments* (1974), *American federalism* (Elazar 1972), and *Quality-of-life indicators in US metropolitan areas* (Liu 1976). While there is some overlap among Liu's life-quality indicators and between them and several metropolitan characteristics analyzed in this study,[4] this bias is assumed not to be significant given the large number of indicators included in the Liu index, their transformation and reduction to adjusted scores, and the further reduction of these measures by factor analysis.

Results

To test the hypothesis of racial dualism, I correlated the overall quality of life with four representative indicators: (a) percent black, (b) percent Spanish-speaking, (c) the black male : total male unemployment rate, adjusted for education, and (d) the percentage of families with an income below the poverty level or greater than $15,000 in 1970 real dollars.

The results show that regardless of population size, the overall quality of life is significantly lower in metropolitan areas with high concentrations of black residents. The correlation coefficients are moderately strong (−0.5) and all are significant at the 0.001 level. By comparison, the overall quality of life is unrelated to the concentration of Spanish-speaking residents. Neither is there any consistent relationship between the overall quality of life and the remaining two indicators of racial and social inequality.

To assess the effects of economic development, I correlated eight representative indicators with the overall quality of life and then controlled them for differences in racial concentration, i.e. percent black. The eight developmental indicators include: (a) Liu's (1976) index of economic concentration, (b) the value added per worker in manufacturing, in thousands of dollars, (c) income per capita, (d) bank deposits per capita, (e) the percentage of persons 25 and older who have completed four years of high school or more, (f) the median housing value, (g) local government taxes per capita, and (h) local government spending per capita. All of the monetary measures were adjusted for metropolitan differences in the cost of living.

Contrary to industrial theories of economic development, the quality of metropolitan life does not appear to depend on the presence of a diversified economic base or manufacturing productivity. Rather, the major developmental determinants of life quality are a high level of educational attainment and personal income. Their impact continues to hold uniformly steady, regardless of variations in racial concentration. The best predictor of life quality across all size categories is the percentage of adults who have achieved at least a high-school education. Correlation and partial correlation values range from a low of 0.65 to a high of 0.85. Interestingly, educational attainment grows in predictive importance as metropolitan size increases. By contrast, personal income becomes more important as metropolitan size decreases.

To evaluate the cultural hypothesis, I analyzed dummy variables for each of Gastil's (1975) cultural regions, his index of southernness, the 1968 presidential vote for George Wallace, an index of political moralism derived from Elazar's (1972) theory, voting turnout in the 1972 presidential election, and the local tax and spending burdens assumed by metropolitan residents. Once again correlation analysis was used to assess the effects of these indicators on the overall quality of life. Partial correlation analysis was used to determine the degree of association that remains after controlling for differences in racial composition and educational attainment.

The results for the regional indicators tend to identify two distinctive cultural regions in the US – the Upper Midwest and the South. The data also lend support to the hypothesis that northern metroplitan areas offer a quality of life superior to that in southern metropolitan areas. This latter result is perhaps most clearly demonstrated by the moderate to moderately high negative correlation coefficients for Gastil's (1975) index of southernness (−0.38 to −0.61) and the 1968 Wallace vote (−0.55 to −0.68). As amply demonstrated elsewhere (Phillips 1969, p. 28), the appeal of the Dixiecrat presidential candidate decreased markedly along centripetal North–South gradiants centred on the Deep South.

While much of the South's lower life quality may stem from its cultural backwardness in race and education, the life quality advantages of the Upper Midwest appear to go beyond these issues. These additional factors may include, but are not necessarily restricted to, a tradition of social and political moderation (including a rejection of racial demagogues), a predominantly moralistic political culture, a politically active citizenry, and a generalized commitment to strong local government (Gastil 1975, pp. 204–15). As the results for the remaining cultural indicators show, these factors account for significant differences in life quality across all metropolitan areas as well. Moreover, these differences persist even in the face of statistical controls for differences in race and education. Overall the best predictor of metropolitan life quality differences is an index of political moralism based on Elazar's theory.

While the foregoing results demonstrate the separate effects of racial dualism, economic development, and political culture on the overall quality of life, they do not tell us which factors predominate or their relative influence. In order to assess this issue, I regressed the overall quality of life on the strongest life-quality correlates of racial dualism, economic development, and political culture; i.e. percent black, percent high-school educated, and the Elazar index of political moralism. The relative magnitudes of the standardized regression coefficients (beta weights) imply that differences in racial concentration (−0.14 to −0.19) are not as important in explaining metropolitan life-quality differences as differences in educational attainment (0.48 to 0.71) and political culture (0.19 to 0.39). For large metropolitan areas, the primary determinant of life-quality differences is educational attainment. For medium and small metropolitan areas a high quality of life appears to depend both on a uniformly high level of educational attainment and a predominantly moralistic culture. Collectively, the three indicators constitute a relatively good fit with the data, explaining from 65% to 79% of the variation in the overall quality of metropolitan life.

Lessons for the USA and the Third World

Throughout this analysis, I have relied primarily on research and data from the United States. But if the US is in fact a developing country, then the results may also be relevant for less-developed countries, i.e. the Third World. One must be careful, of course, in extrapolating findings from one cultural milieu to another. This is because factors that are assumed constant in one setting, such as democratic government, may be highly variable or even nonexistent in other settings. Notwithstanding, there are several lessons that may apply with equal force to the US and the Third World.

One concerns the social costs of racial dualism. In general most studies of dualism have emphasized its negative consequences for various minority groups. The results of this study, however, show that the effects of dualism are felt by everyone. This is because of the moderately strong and negative correlation found between the concentration of black residents in metropolitan areas and the overall quality of life. Thus the presence of historically excluded groups may tend, with certain exceptions (such as the Spanish-speaking), to lower the quality of life for all residents.

A second lesson relates to the priority of human development over purely economic development. The results show that the key to a superior quality of life, at least in the US, is not a diversified economy of manufacturing productivity but the development of an educated middle class. This social stratum also seems critical to the formation of stable communities and the provision of public services (including education) that help sustain them. This inference is supported by the moderate to strong positive correlations found between the overall quality of metropolitan life and the median housing value, taxes per capita, and spending per capita.

A third lesson involves the independent contributions of political culture to the overall quality of life. All things equal, this study finds that moralistic (participant) cultures do the most to enhance the overall quality of life, individualistic (commercial) cultures do the next most, and traditionalistic (parochial) cultures do the least. In the US moralistic cultures are generally concentrated in the northern tier of states – New England, the Upper Midwest, and the Pacific Northwest.[5] Racially they are more homogeneous and more heavily populated by people of northern European extraction; economically they are more self-sufficient; and socially they are more cohesive and law-abiding. Politically, moralistic cultures tend to be more progressive; i.e. more intolerant of political corruption, more likely to register and vote, and more likely to support higher levels of public services (including education) and higher tax burdens.

Perhaps a final lesson pertains to the significant and cumulative impact of race, education, and culture on the quality of American life. Though their separate effects have been statistically isolated for analytical purposes, it is clear that the three are inextricably bound together. Thus the metropolitan areas and regions with the highest overall quality-of-life scores are those that are least beset by problems of racial dualism, most uniformly educated, and most moralistic. To conclude, the quality of life in US metropolitan areas depends fundamentally on the quality of the people who inhabit them.

Notes

1. This reversal of past trends is apparently due to the more rapid breakdown of the black family (Auletta 1981).
2. Witness for example the recent decisions by Nissan Motors and General Motors in selecting metropolitan Nashville as the site for new auto plants.
3. It is perhaps no accident that as of this writing, at least seven American cities have passed resolutions offering sanctuary and support for Central American refugees (Cleveland *Plain Dealer* 1986).
4. Per capita income for example is one of 123 indicators included in Liu's index and also one of the selected indicators of economic development.
5. The Mormon region also appears to be dominated by a moralistic culture (Elazar 1972, Gastil 1975, Lieske 1984).

References

Almond, G. & S. Verba 1963. *The civic culture*. Boston: Little Brown.
Auletta, K. 1981. *The underclass*. New York: Random House.
Bartley, N. V. & H. D. Graham 1975. *Southern politics and the second reconstruction*. Baltimore: Johns Hopkins University Press.
Boyer, R. & D. Savageau 1985. *Places rated almanac*. Chicago: Rand McNally.
Cleveland *Plain Dealer* 1986. Seattle welcomes Latin refugees. January 14.
Doeringer, P. & M. J. Piore 1975. Unemployment and the dual labor market. *Public Interest* **38**, 67–79.
Elazar, D. 1970. *Cities of the prairie*. New York: Basic Books.
Elazar, D. 1972. *American federalism*, 2nd edn. New York: Thomas Crowell.
Elgin, D., T. Thomas, T. Logothetti & S. Cox 1974. *City size and the quality of life*. Menlo Park, Cal.: Stanford Research Institute.
Erikson, R. S., N. R. Luttbeg & K. L. Tedin 1980. *American Public Opinion*, 2nd edn. New York: Wiley.
Fallows, J. 1983. The New Immigrants. *Atlantic* (November): 45–106.
Gastil, R. D. 1975. *Cultural regions of the United States*. Seattle: University of Washington Press.
Key, V. O., Jr. 1949. *Southern politics*. New York: Vintage.
Kincaid, J. 1980. Political culture and the quality of urban life. *Publius* **10** (Spring): 89–110.
Lieske, J. 1984. The salvation of American cities. In P. Porter & D. Sweet (eds.) *Rebuilding America's cities*. New Brunswick, N.J.: Center for Urban Policy Research.
Liu, Ben-Chieh 1976. *Quality-of-life indicators in US metropolitan areas*. New York: Praeger.
Marlin, J. T., I. Ness & S. T. Collins 1986. *Book of world city rankings*. New York: Free Press.
Pear, R. 1982. Inflation wiped out gains in earnings in 70s. *New York Times*, April 25.
Perry, D. & A. Watkins 1977. *The rise of the sunbelt cities*. Beverly Hill, Cal.: Sage.
Pettigrew, T. & R. Spier 1962. The ecological structure of Negro homicide. *American Journal of Sociology* **67**, 621–9.
Phillips, K. P. 1969. *The emerging Republican majority*. New Rochelle, N.Y.: Arlington House.
Reich, M. 1977. The economics of racism. In *Problems in Political Economy*, 2nd edn, D. M. Gordon (ed.). Lexington, Mass.: D. C. Heath.
Sharkansky, I. 1970. *Regionalism in American politics*. New York: Bobbs-Merrill.
Sharkansky, I. 1975. *The United States: a study of a developing country*. New York: David McKay.
Sivard, R. L. 1983. *World military and social expenditures*. Washington, DC: World Priorities.
Smith, D. 1973. *The geography of social wellbeing*. New York: McGraw-Hill.
US Bureau of the Census 1973. *Census of the population, 1970*. Washington: Government Printing Office.
US Bureau of the Census 1974. *Census of governments, 1972*, Vol. 5. Washington: Government Printing Office.
Verba, S. & N. H. Nie 1967. *Participation in America*. New York: Harper & Row.
Weinstein, B. & R. Firestine 1978. *Regional growth and decline in the United States*. New York: Praeger.
Winham, G. R. 1970. Political development and Lerner's theory. *American Political Science Review* **64**, 810–18.
Zelinsky, W. 1973. *The cultural geography of the United States*. Englewood Cliffs, NJ: Prentice-Hall.

Afterword

JIM NORWINE

> It has always been a mystery to me how men can feel themselves honored by the humiliation of their fellow-beings.
>
> M. K. Gandhi 1927: *Gandhi: an autobiography*

I originally conceived this book as a primer for readers in the First World, although I did hope that Third World readers would also find it worthwhile. Thanks to its contributors, however, *The Third World* evolved into the much broader work it is. A brief comment concerning that evolution might be in order.

During the early years of the 1980s, I became intrigued (obsessed?) with the question of (a) whether most educated residents of the developed world had the developing world on their global "mental maps" at all, and (b), if so, just how accurate was their intellectual cartography? The answers I felt I found were: (a) yes, to a limited extent, but (b) not very – accurate, that is.

It seemed clear even then that imperious First World ignorance about the lives of two-thirds of humanity was anything but bliss. Today, with the USA embroiled in bitter disputes with Syria, Iran, and Libya (and, indirectly, at least to some modest degree, with much of the non-aligned world in general), such geographic illiteracy is at best fatuously myopic and at worst suicidally dangerous.

The Third and First Worlds are *different*: comprehension of this simple but pre-eminent fact represents the foundation upon which more deeper insights may be built. Grasping this first principle, although only a beginning, is a requisite if more profound enlightenment is to be achieved. (This is, of course, equally true for First *and* Third World readers, which is why the scope of this book became much more general than I had first anticipated.)

It is imperative to understand that the key question regarding First/Third World differences is not, as is often assumed, "Is life better in the First World or in the Third World?" The answer to that is too easy: "yes." Which is, of course, merely a facile way of reminding ourselves that things that are different in kind – e.g., an apple and an orange – cannot be qualitatively compared without some reasonable sense of the nature of each (e.g., apples are crunchier, but oranges juicier).

As everyone knows, the First/Second and Third Worlds (usually) differ in degree of affluence, the developed world having greater wealth. However, it is not unlikely that they differ even more in perspectives, attitudes, values, and cultural traditions. Consider this telling passage from Jean Raspail's *The camp of the saints*:

> The West doesn't like to burn its dead. It tucks away its cremation urns, hides them out in the hinterlands of its cemetaries. The Seine, the Rhine, the Loire, the Rhône, the Thames are no Ganges or Indus . . . Their shores never stank with the stench of roasting corpses. Yes, they have flowed with blood, their waters have run red, and many a peasant has crossed himself as he used his pitchfork to push aside the human carcasses floating downstream. But in Western times, on their bridges and banks, people danced and drank their wine and beer, men tickled the fresh, young, laughing lasses, and everyone laughed at the wretch on the rack, laughed in his face, and the wretch on the gallows, tongue dangling, and the wretch on the block, neck severed – because, indeed, the Western World, staid as it was, knew how to laugh as well as to cry – and then, as their belfrys called them to prayer, they would all go partake of their fleshly god, secure in the knowledge that their dead were there, protecting them, safe as could be, laid out in rows beneath their timeless slabs and crosses, in graveyards nestled against the hills, since burning, after all, was only for devilish fiends, or wizards, or poor souls with the plague . . . (Raspail 1973)

Just as the Third World has its serf-like women and children, its illiteracy, its *physical* despair, the First World suffers its own peculiar despair which, although partly physical (e.g., cardiovascular diseases and cancers), is primarily the psychological, intellectual or *spiritual* "disease" (as Mother Teresa puts it) of alienation and loneliness. Dr. Benjamin Spock, medical guru to the American "yuppie" or postwar baby-boom generation, has rather disgustedly observed that, ". . . divorce rates have doubled. Teen-age suicide rates have tripled. We don't have spiritual beliefs. We don't have souls anymore." (Spock 1986.)

I shan't belabor this point, merely recall the good Mother's metaphoral injunction that although the Third World needs the First to save its "body", the First needs the Third in order to save its "soul" . . . . Can we ever hope to achieve such an idealistic goal? Perhaps, perhaps not. In any event, appreciation and understanding of one another clearly represents the first step toward its attainment.

I hope that *The Third World* has provided a conceptual framework, an embarcation point, for its readers. It could do no more than that. Our goal was to produce a well-written, enlightening and current collage which, while answering many questions, would also whet our readers' appetites for more. Deeper works than this exist for those interested in studying specialized topics.

The Third World, as I hope this work makes clear, is as much *experience* as it is conceptual or even geographic realm. It must therefore be experienced first-hand if it is to be fully appreciated.

A very useful stop-gap measure is reading those "social commentaries" – essays and books – of writers who are cogent observers and students of their own and other cultures. Particularly helpful is the work of a small coterie of essayists who combine delightfully readable – if generally irreverent – prose with thoughtful, contemplative observation. An abbreviated listing, for instance, should include the non-fiction of Paul Theroux; Lawrence Durrell; V. S. Naipaul (e.g., *India: a wounded civilization*); Shiva Naipaul ("The myth of the Third World," *The Spectator*, May 18, 1985); William Golding (*An Egyptian journal*); Yi-Fu Tuan (e.g., *The good life*); James Michener (e.g., *Iberia*); Warren Johnson (*Muddling toward frugality*); Luigi Barzini (*The Italians*); Octavio Paz (*The labyrinth of solitude* or *One earth, four or five worlds*); German Ariniegas (*America in Europe*); Nigel Barley (*Adventures in a mud hut*); Steven W. Mosley (*Journey to the forbidden China*); Robert Repetto (*The global possible*); Elizabeth and Robert Fernea (*The Arab world: personal encounters*); Jean Raspail (*The camp of the saints*); and those additional works presented in the list of references and further reading.

For the reader who still doubts that the Third World is both states of mind and of being, I presume to offer an unusual suggestion in the form of several popular motion pictures and novels. Perhaps this is not so surprising after all, for art is indeed "a lie which reveals the truth", the very sort of oblique perspective which, as we noted at the beginning of our Introduction, seems vital to understanding the Third World. These recent novels stand out as unusually revealing with respect to Third World states of mind, and they are all extremely readable: *The year of the French*, by Thomas Flanagan; *The city of joy*, by Dominique Lapierre; Jean Raspail's *The camp of the saints* and Umberto Eco's *The name of the rose*. The films are *El Norte*, *The gods must be crazy*, and *Lost in America*.

Ultimately, however, experiencing the Third World requires personal journeying in Bhutan or Bolivia, Burundi or Bangladesh, with eye, mind, and itinerary wide open. Likewise, Third World citizens must be encouraged – and assisted – to experience personally the First World. I will count this effort a success if it stimulates some readers to leave their armchairs and do just that.

References and further reading

Abbott, E. A. 1928. *Flatland*. Boston: Little Brown.

Annual Editions 1986. *Global issues 86/87*. Guilford, Conn.: The Dushkin Publishing Group.

Antipode: A Radical Journal of Geography. P.O. Box 339, West Side Station, Worcester, Mass. 01602.

Ardniegas, G. 1986. *America in Europe*. Orlando, Fla.: Harcourt Brace Jovanovich.

Bailey, N. 1984. *Adventures in a mud hut*. New York: Vanguard.

Barke, M. & G. O'Hare 1984. *The Third World. Diversity, change and interdependence*. Edinburgh: Oliver & Boyd.

Barzini, L. 1964. *The Italians*. New York: Bantam Books.

Boulding, K. 1980. "Science: our common heritage". Presidential address, AAAS Annual Meeting, January 1980, San Francisco, California.

Brody, H. 1982. *Maps and dreams*. New York: Pantheon Books.

Brown, L. R. *et al*. 1986. *State of the world: 1986*. New York: Norton.

Capra, F. 1984. *The Tao of physics*. New York: Bantam Books.

Carson, R. 1965. *The sense of wonder*. New York: Harper & Row.

Chaffetz, D. 1981. *A journey in Afghanistan*. Chicago: University of Chicago Press.

Chatwin, B. 1977. *In Patagonia*. New York: Summit Books.

Chatwin, B. 1987. *The songlines*. New York: Viking.

Cole, J. P. 1983. *Geography of world affairs*. Norwich: Butterworths.

Crane, S. 1892. "The broken down van," in *The New York Tribune*. See Angus Paul, *The Chronicle of Higher Education*, January 13, 1987. V. 34, No. 18.

Crow, B. & A. Thomas 1983. *Third World atlas*. Philadelphia: Milton Keynes.

Davenport, G. *The geography of the imagination*. San Francisco: North Point Press.

Davenport. G. 1987. *Every force evolves a form*. San Francisco: North Point Press.

de Poncins, G. 1941. *Kabloona*. New York: Reynal.

de Poncins, G. 1957. *From a Chinese city*. New York: Doubleday.

Dickenson, J. P. *et al*. 1983. *A geography of the Third World*. London: Methuen.

Dreiske, N. 1986. "Macondo" offers Americans portrait of a different South America. News release, Facets Multimedia Inc., 1517 West Fullerton Ave., Chicago, Ill. 80614.

Duggan, W. 1985. *The great thirst*. New York: Delacorte Press.

Eco, U. 1983. *The name of the rose*. New York: Harcourt Brace Jovanovich.

Ehrlich, G. 1985. *The solace of open spaces*. New York: Viking Press.

Eliade, M. 1969. *The quest*. Chicago: University of Chicago Press.

Eliot, G. 1872. *Middlemarch*. New York: E. P. Dutton (1930).

Fermor, P. L. 1977. *A time of gifts*. New York: Harper & Row.

Fermor, P. L. 1986. *Between the woods and water*. New York: Viking.

Fernea, E. W. & R. A. Fernea 1985. *The Arab World: personal encounters*. Garden City, NY: Anchor Press.

Flanagan, T. 1979. *The year of the French*. New York: Pocket Books.

Flanagan, T. 1988. *The tenants of time*. New York: E. P. Dutton.

Fleming, P. 1933. *Brazilian adventure*. New York: Charles Scribner's Sons.

Frank, A. G. 1981. *Crisis: In the Third World*. New York: Holmes & Meier.

Gandhi, M. K. 1957. *Gandhi: an autobiography*. Boston: Beacon Press.

García Márquez, G. 1971. *One hundred years of solitude*. New York: Avon.

Goldberg, N. 1986. *Writing down the bones*. Boston: Shambala Publications.

Golding, W. 1985. *An Egyptian journal*. Winchester, Mass.: Faber & Faber.

Granta: Vol. 10, *Travel writing* (1984); Vol. 21, *The story-teller* (1987); Vol. 22, *With your tongue down my throat* (1987). Harmondsworth: Granta Publications.

Hardin, G. & J. Baden 1977. *Managing the commons*. San Francisco: Freeman.

Harrer, H. 1953. *Seven years in Tibet*. New York: Dutton.
Hesse, H. 1957. *The journey to the east*. New York: The Noonday Press.
Hesse, H. 1951. *Siddhartha*. New York: New Directions.
Holmes, R. 1984. "In Stevenson's footsteps", in *Granta*, Vol. 10: *Travel writing*. Harmondsworth: Granta Publications.
Hudson, W. H. 1939. *Far away and long ago*. London: J. M. Dent.
Hunger Project, The. 1985. *Ending hunger: an idea whose time has come*. New York: Praeger Publishers.
Jennings, G. 1980. *Aztec*. New York: Avon Books.
Johnson, W. 1978. *Muddling toward frugality*. Boulder, Col.: Shambhala Publications.
Jordan, T. G. & L. Rowntree. 1986. *The human mosaic: a thematic introduction to cultural geography*. New York: Harper & Row.
Khan, H. I. 1978. *The complete sayings of Hazrat Inayat Khan*. New Lebanon, NY: Sufi Order Publications.
Kurian, G. T. 1979. *Encyclopedia of the Third World*. London, U.K.: Mansell.
Levi, P. 1987. *The light garden of the angel king*. London: Penguin.
Michener, J. A. 1968. *Iberia: Spanish travels and reflections*. Greenwich, Conn.: Fawcett.
Morris, J. 1985. *Last letters from Hav*. New York: Random House.
Morris, J. 1980. *Destinations*. New York: Oxford.
Mosher, S. W. 1985. *Journey to the forbidden China*. New York: Free Press.
Munro, E. 1987. *On glory roads*. New York: Thames & Hudson.
Naipaul, S. 1980a. *Journey to nowhere: a new world tragedy*. New York: Simon & Schuster.
Naipaul, S. 1980b. *North of South: an African journey*. New York: Penguin Books.
Naipaul, S. 1985a. *Beyond the dragon's mouth*. New York: Viking.
Naipaul, S. 1985b. A thousand million invisible men: the myth of the Third World. *The Spectator*, 18 May, 9–11.
Naipaul, V. S. 1977. *India: a wounded civilization*. New York: Knopf.
Naipaul, V. S. 1981. *Among the believers: an Islamic journey*. New York: Knopf.
Naipaul, V. S. 1984. *Finding the center*. New York: Knopf.
Newby, E. 1968. *A short walk in the Hindu Kush*. Harmondsworth: Penguin.
New Internationalist: the people, the ideas, the action in the fight for world development. 42 Hythe Bridge Street, Oxford OX1 2EP, UK. (Monthly journal.)
Norwine, J. 1978. *Climate and human ecology*. Houston: Armstrong.
Norwine, J. 1980. *Geography as human ecology*. Washington, DC: University Press.
Parkman, F. Jr. 1849. *The Oregon trail*. Madison, Wis.: University of Wisconsin Press (1969).
Paz, O. 1961. *The labyrinth of solitude: life and thought in Mexico*. New York: Grove Press.
Paz, O. 1984. *One Earth, four or five worlds*. San Diego: Harcourt Brace Jovanovich.
Pirages, D. 1978. *Global ecopolitics: the new context for international relations*. North Scituate, Mass.: Duxbury.
Pirsig, R. M. 1974. *Zen and the art of motorcycle maintainance or an inquiry into values*. New York: Morrow.
Population Reference Bureau, Inc. Population Bulletins (various). 2213 M Street N.W., Washington, DC 20037.
Raspail, J. 1975. *The camp of the saints*. New York: Ace Books.
Reid, A. 1987. *Whereabouts: notes on being a foreigner*. San Francisco: North Point Press.
Reitsma, H. A. & J. M. G. Kleinpenning 1985. *The Third World in perspective*. Totowa, NJ: Rowman & Allanheld.

Repetto, R. (ed.) 1985. *The global possible*. New Haven, Conn.: Yale University Press.
Resources. (Journal published four times per year.) Resources for the Future, Inc., 1616 P Street N.W., Washington, DC 20036.
Rucker, R. 1983. *The 57th Franz Kafka*. New York: Ace Books.
Rutherfurd, E. 1987. *Sarum*. New York: Crown.
Simeti, M. T. 1986. *On Persephone's island*. New York: Knopf.
Smith, D. M. 1979. *Where the grass is greener: living in an unequal world*. Baltimore: The Johns Hopkins University Press.
Spock, B. 1986. Quoted in: "Baby boomer bashing: is it America's favorite new sport?", by Lynnell Mickelsen, May 25, Section J, page 1, *Corpus Christi Caller-Times*, Corpus Christi, Tx.: Caller-Times.
Steegmuller, F. 1972. *Flaubert in Egypt: a sensibility on tour*. Boston: Little Brown.
Third World Forum. Association of Third World Affairs, 1712 Corcoran St. N.W., Washington, DC 20009. (Bi-monthly journal.)
Todaro, M. P. 1985. *Economic development in the Third World*. New York: Longman.
Transition. (Quarterly journal of the Socially and Ecologically Responsible Geographers.) Cincinnati: Dept. of Geography, University of Cincinnati, Ohio 45221.
Trollope, F. 1949. *Domestic manners of the Americans*. New York: Knopf.
Tuan, Yi-Fu 1986. *The good life*. Madison: The University of Wisconsin Press.
UFSI reports. (Periodic in-depth country/topical reports.) Universities Field Staff International, Inc., 620 Union Drive, Indianapolis, Ind. 46202.
Volgeler, Ingolf & A. R. de Souza 1980. *Dialectics of Third World development*. Totowa, NJ: Allanheld, Osmun.
Wheeler, J. 1973. In J. Mehra, *The Physicist's Conception of Nature*. Dordrecht, Holland: D. Reidel.
World development forum. (Bi-monthly newsletter.) The Hunger Project, 1717 Massachusetts Ave. N.W., Suite 604, Washington, DC 20036.
World Eagle. (Monthly journal.) World Eagle, Inc., 64 Washburn Ave., Wellesley, Mass. 02181.
World Resources Institute 1986a. *Journal '86: the annual report of the World Resources Institute*. Holmes, Pa.: WRI Publications.
World Resources Institute 1986b. *World resources 1986*. New York: Basic Books.
Zuckmayer, C. 1970. *A part of myself*. London: Secker & Warburg.

Contributors

Thomas D. Anderson, Department of Geography, Bowling Green State University, Ohio, USA

Alice C. Andrews, Department of Geography, George Mason University, Fairfax, Virginia, USA

Robert S. Bednarz, Department of Geography, Texas A & M University, USA

John P. Cole, Department of Geography, University of Nottingham, UK

Richard J. Estes, School of Social Work, University of Pennsylvania, USA

Jerome D. Fellman, Department of Geography, University of Illinois, USA

Hal Fisher, Departments of Journalism and Mass Communications, Bowling Green State University, Bowling Green, Ohio, USA

John R. Giardino, Departments of Geography and Geology, Texas A & M University, USA

Alfonso Gonzalez, Department of Geography, University of Calgary, Canada.

Gerald L. Ingalls, Department of Geography, University of North Carolina, USA

Raja Kamal, Director, Middle East Fellowship Program, Brandeis University, Boston, USA

Joel Lieske, Department of Political Science, Cleveland State University, Ohio, USA

Walter E. Martin, Department of Geography, University of North Carolina, USA

Allen H. Merriam, Missouri Southern State College, Joplin, Missouri, USA

Jim Norwine, Department of Geosciences, Texas A & I University, Kingsville, Texas, USA

Evelyn Peters, Department of Geography, Queen's University, Kingston, Ontario, Canada

B. L. Sukhwal, Department of Geosciences, University of Wisconsin-Platteville, USA

Laurence Grambow Wolf, Department of Geography, University of Cincinnati, USA

Yi-Fu Tuan, Department of Geography, University of Wisconsin, Madison, USA

Index

Italic numbers denote Figures in text.